International Conference on Innovative Research (ICIR Euroinvent 2022)

IOP Conference Series: Materials Science and Engineering
Volume 1251

Iasi, Romania
25 - 27 May 2022

Editors:

Andrei Victor Sandu
Mohd Mustafa Al Bakri Abdullah
Petrica Vizureanu
Marcin Nabialek
Che Mohd Ruzaidi Ghazali
Ion Sandu

ISBN: 978-1-7138-6017-4
ISSN: 1757-8981

Printed from e-media with permission by:

Curran Associates, Inc.
57 Morehouse Lane
Red Hook, NY 12571

Some format issues inherent in the e-media version may also appear in this print version.

Institute of Physics
Dirac House, Temple Back
Bristol BS1 6BE UK

Phone: 44 1 17 929 7481
Fax: 44 1 17 920 0979

techtracking@iop.org

Additional copies of this publication are available from:

Curran Associates, Inc.
57 Morehouse Lane
Red Hook, NY 12571 USA
Phone: 845-758-0400
Fax: 845-758-2633
Email: curran@proceedings.com
Web: www.proceedings.com

TABLE OF CONTENTS

ICIR Euroinvent 2022

International Conference on Innovative Research

in the field of Materials Science and Engineering

May 26th to 27th, 2022
Iasi – Romania

Organized by:

- **Romanian Inventors Forum**
- **Faculty of Materials Science and Engineering, The "Gheorghe Asachi" Technical University of Iasi, Romania**
- **ARHEOINVEST Platform, Alexandru Ioan Cuza University of Iasi**
- **Centre of Excellence Geopolymer and Green Technology CEGeoGTech), Universiti Malaysia Perlis (UniMAP)**
- **Department of Physics, Czestochowa University of Technology, Częstochowa, Poland**

With support of:

- **Universiti Malaysia Terengganu**
- **International Federation of Inventors' Associations – IFIA**
- **World Invention Intellectual Property Associations – WIIPA**

Editors:
Andrei Victor SANDU, Mohd Mustafa Al Bakri ABDULLAH, Petrică VIZUREANU, Marcin NABIALEK, Che Mohd Ruzaidi GHAZALI, Ion SANDU

IOP Conf. Series: Materials Science and Engineering 1251 (2022) 011001 doi:10.1088/1757-899X/1251/1/011001

Foreword

This volume contains selected peer-reviewed articles presented at ICIR Euroinvent ICIR Euroinvent 2021 Conference. The event was held on-site with on-line transmission in Iaşi, România from 26th to 27th of May 2022.

The ICIR Conference is organized under the auspices of EUROINVENT. Euroinvent is a joint event promoting creativity in a European context, by displaying the contributions of consecrated schools from higher education and academic research and also of individual inventors and researchers.

The EUROINVENT International Conference on Innovative Research (ICIR) brings together leading researchers, engineers and scientists who will present actual research results in the field of Materials Science and Engineering.

The conference aims to provide a high level international forum for researchers, engineers and scientists to present their new advances and research results in the field of materials science and engineering.

The volume covers all the aspects of materials science, from synthesis and characterization of materials to procedures and technologies for materials engineering, as well as materials application and their involvement in the life sciences.

All the papers have been reviewed by at least two expert referees in their relevant topic disciplines. The papers selected for the volume depended on their quality and relevancy to the conference.

The structure of the conference was kept similar to the previous ones, having 3 Keynote Speakers (renowned scientists from Romania – Prof. Florin MICULESCU, Portugal – Prof. Abílio P. SILVA and Malaysia – Prof. Mohd Mustafa al Bakri ABDULLAH) who had 30 minutes to present their research and to reply to questions, also 4 Invited Speakers (Ukraine – Prof. Yulia IVASHKO, Poland – Dr. Mariusz WINIECKI and Dr. Wojciech SOCHACKI and from Romania– Prof. Lidia BENEA), who had 20 minutes to present their research and to reply to questions. To the regular presenters were given 15 minutes for presentation and Q&A. In total were 37 plennary presentation next to 50 posters. From 122 registrations with 101 registered abstracts, only 87 were accepted futher for presentation and only 18 of them were accepted for publication to IOP MSE.

The conference was very dynamic with many questions and replies from the participants. During the sessions more than 200 participants joined the presentations from Bulgaria, Hungary, Israel, Japan, Malaysia, Moldova, Morroco, Poland, Romania, Spain, Thailand, Indonesia, Ukraine and not only.

At conference closure ceremony, on the decision of the Scientific Board, Two Best Oral Presentation Award was presented (2 Awards of 500 EUR each provided by MDPI Materials and Micromachines Journals), next to 2 Best Poster Awards. These awards where handled personally by His Excellency Minister of Research, Education and Digitization of Romania Mr. Sebastian BURDUJA and Mayor of Iasi city Mr. Mihai CHIRICA.

The editors hope that this volume will provide the reader a broad overview of the latest advances in the field of materials science and engineering and that they will be a valuable references source for further research.

The editors would like to express their sincere appreciations and thanks to all the committee members of the ICIR 2022 for their tremendous efforts. Thanks also to IOP for producing the volume with full articles.

Finally, the editors would like to thank all the authors for their contribution to this valuable volume.

Editors

IOP Conf. Series: Materials Science and Engineering 1251 (2022) 011001 doi:10.1088/1757-899X/1251/1/011001

EUROINVENT ICIR 2022 — International Conference on Innovative Research, Iasi, 26th–27th of May 2022

Scientific Advisory Board

Prof.Dr. Ion SANDU
Alexandru Ioan Cuza University of Iasi, Romania – President
Prof.Dr. Mohd Mustafa Al Bakri ABDULLAH,
Malaysian Research & Innovation Society, Malaysia
Prof.Dr. Iulian ANTONIAC
Politehnica University Bucharest, Romania
Prof.Dr. Lidia BENEA
Dunarea de Jos University of Galati, Romania
Prof.Dr. Katarzyna BLOCH
Czestochowa University of Technology, Poland
Prof.Dr. Kosta BOSHNAKOV
University of Chemical Technology and Metallurgy, Sofia, Bulgaria
Prof.Dr. Antonio CANDEIAS
University of Evora, Portugal
Prof.Dr.Eng. Dan CASCAVAL
Gheorghe Asachi Technical University of Iasi, Romania
Prof.Dr. Christian CODDET
Universite de Technologie Belfort-Montbeliard, France
Prof.Dr. Che Mohd Ruzaidi GHAZALI
Universiti Malaysia Terengganu, Malaysia
Prof.Dr. Kamarudin HUSSIN
Universiti Malaysia Perlis, Malaysia
Prof.Dr. Marcin NABIALEK
Czestochowa University of Technology Poland
Prof.Dr.Eng. Catalin POPA
Technical University of Cluj-Napoca, Romania
Prof.Dr.Eng. Cristian PREDESCU
Politehnica University Bucharest, Romania
Prof.Dr.Eng. Cornel SAMOILA
Transilvania University of Brasov, Romania
Prof.Dr.Eng. Neculai Eugen SEGHEDIN
Gheorghe Asachi Technical University of Iasi, Romania
Prof.Dr.Eng. Mihail Aurel TITU
Lucian Blaga University of Sibiu, Romania
Prof.Dr. Jerzy J. WYSLOCKI
Czestochowa University of Technology, Poland

Chairman

Prof. Dr. Eng. Petrica VIZUREANU
The "Gheorghe Asachi" Technical University of Iasi, Romania

Event Coordinator

Assist.Prof.Dr.Eng. Andrei Victor SANDU
Romanian Inventors Forum &
The "Gheorghe Asachi" Technical University of Iasi, Romania

IOP Conf. Series: Materials Science and Engineering 1251 (2022) 011001 doi:10.1088/1757-899X/1251/1/011001

EUROINVENT ICIR 2022 — International Conference on Innovative Research, Iasi, 26th–27th of May 2022

Technical and Organizing Committee

IOP Conf. Series: Materials Science and Engineering 1251 (2022) 011002 doi:10.1088/1757-899X/1251/1/011002

Peer Review Statement

All papers published in this volume have been reviewed through processes administered by the Editors. Reviews were conducted by expert referees to the professional and scientific standards expected of a proceedings journal published by IOP Publishing Publishing.

- **Type of peer review:** Single Anonymous
- **Conference submission management system:** All via email: Professor Andrei Victor Sandu sav@tuiasi.ro
- **Number of submissions received:** 28
- **Number of submissions sent for review:** 28
- **Number of submissions accepted:** 18
- **Acceptance Rate (Submissions Accepted / Submissions Received × 100):** 64.3
- **Average number of reviews per paper:** 2
- **Total number of reviewers involved:** 26
- **Contact person for queries:**
  Name: Sandu Andrei Victor
  Email: sav@tuiasi.ro
  Affiliation: Gheorghe Asachi Technical University of Iasi

IOP Conf. Series: Materials Science and Engineering 1251 (2022) 012001 doi:10.1088/1757-899X/1251/1/012001

Overview on plasma electrolytic oxidation of magnesium alloys for medical and engineering applications

I Patrascu[1,2], M C Ducu[1,3*], A D Negrea[3], S G Moga[3] and A G Plaiasu[1]

[1]University of Pitesti, Manufacturing and Industrial Management Department, Interdisciplinary Doctoral School,110040, Pitesti, Romania
[2]IP Automatic Design, Industriei nr 6, Lunca Corbului 117435, Arges, Romania
[3]Regional center of research & development for materials, processes and innovative products dedicated to the automotive industry (CRC&D- Auto), 110440, Pitesti, Romania

E-mail: catalin.ducu@upit.ro

Abstract. The scope of this review article is to offer a general view of the recent advances in the plasma electrolytic oxidation of Mg and Mg alloys with emphasis of the Plasma Electrolytic Oxidation (PEO) processing parameters used to obtain key surface properties for medical and engineering applications. The paper presents the different dependencies of the PEO process, first of all depending on the composition of the substrate, the electrolyte used, the applied electrical regime, as well as the way of their interaction, with direct implications on the properties designed to satisfy concrete applications for magnesium alloys. Magnesium-deformable alloys are of particular interest for the manufacture of parts for the for medical and engineering applications and industry due to the possibility of obtaining a more homogeneous structure and better mechanical properties, compared to cast parts. Also, magnesium alloys find its place to a broad range of aerospace, electronics, commercial and sport-related applications. In general, the interactions of the substrate / electrolyte combination with the electrical regime are complex and still the subject of ample research. Because, unlike other electrolytic surface treatment methods, PEO results in the formation of high-strength ohmic layers, they affect, especially in the case of current-controlled regimes, the extent to which the predefined electrical pulse is projected correctly in the setting experimental data.

1. Introduction

Functional surfaces have received special attention in recent decades due to their needs increasingly complex and diverse people, as well as the need to obtain materials with improved properties regardless of the environment in which they are used and which require both a low-price production as well as a low maintenance cost.

The industry requires that the materials used in the manufacture be meet the required specifications of increased resistance to fatigue and corrosion, resistance improved breaking strength and have a high strength / weight ratio due to their service condition. The ability to obtain functional surfaces, that are superior in terms of their mechanical characteristics (hardness, wear resistance, adhesion to substrates metal, fatigue strength), physicochemical properties (resistivity, tensile strength), rupture, thermal shock resistance and heat protection) and corrosion resistance, can be provide it by PEO. The cost and duration of the production process are kept to a minimum because of the preparation (washing, degreasing, pickling, clarification) normally required for the surfaces of the parts and structures to be covered are not required in PEO process. Also the process is quite environmentally friendly because the solution used in it introduces relatively low concentrations of chemical components non-toxic in the ecosystem. In addition, the PEO process is a one-step technique that can be used to treat surfaces with complex 3D surface geometry and a wide range of sizes, which is not possible in the case of many other surface treatment methods. In recent years database Web of Science specifies the researchers' interest, in studying the PEO process for medical and industrial applications shows an increase as in the Figure 1.

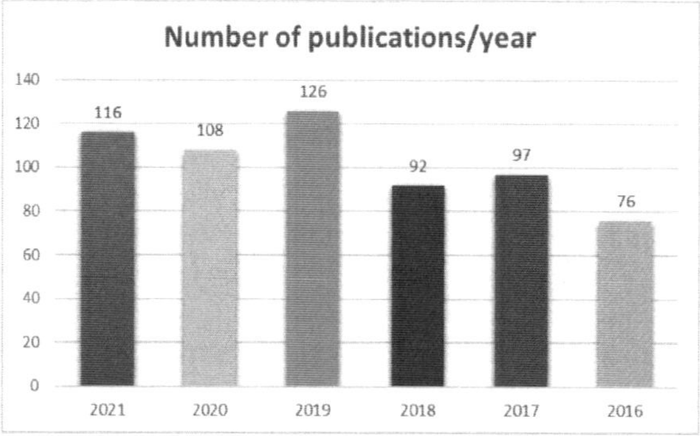

Figure 1. Number of publications in Web of Science.

The tendency to further reduce the weight and assemblies of parts has led to use surface treatments of magnesium and its alloys in key aerospace and automotive industries. Magnesium and its alloys can be also good candidates for medical applications, where biocompatible and bioresorbable materials are needed. Key industrial sectors already benefit from the physical and chemical properties of Mg alloys, like [1]: low density, high specific strength and stiffness, good damping performance, good bio-compatibility, large hydrogen storage capacity, etc. Some advantages of Mg and its alloys vs steel or aluminum are: low density, high recyclability, better ductility, excellent castability and easiness to machine [2].

Resorbable metal materials are the next step in the development of implants absorbable orthopedic agents, Mg and its alloys, in particular calcium alloys, representing an optimal solution for obtaining resorbable implants, having in its composition elements naturally present in the human body. These have the advantage of properties mechanical superior to biopolymers. The density and elastic modulus are close to those of human bone tissue, thus avoiding the effect of the stress shielding effect. To improve the corrosion behavior, with positive effect on biocompatibility of magnesium and its alloys, there are 2 main methods, adjustment microstructure by alloying or machining and performing surface treatments and use of coatings, like PEO process.

2. Plasma electrolytic oxidation of magnesium alloys for medical and engineering applications

2.1. Plasma electrolytic oxidation of Mg and its alloys for bio-medical applications

Magnesium and its alloys are attractive materials for bio-medical applications to their intrinsic properties like: low weight, density close to that of bone and a high strength (tensile) ratio in relation to weight [3]. Much more, they are bioactive materials with resorbable properties. Some short-comings for application of Mg as bio-medical device is its high reactivity (high corrosion rate under physiological conditions and gaseous hydrogen release in surrounding tissues) and the mechanical characteristics [4]. A successful Mg-based bio-material needs to meet the following criteria: biosafety and biocompatibility, good mechanical properties, uniform degradation, cost effective eco-friendly production techniques [5].

One first step for the improvement of the mechanical and electrochemical properties of Mg-based materials, assumes a proper choice of alloying elements in respect with bio-medical requirements. Another strategy is to modify and functionalize the surface of the Mg-based implants by surface coating techniques in order to restrain the corrosion rate and increase the bioavailability of the implants. Plasma electrolytic oxidation is an attractive technique for surface functionalization of Mg-based bio-medical devices. It is an environmentally friendly process that produces conversion layers with good adhesion and biocompatibility. Judicious choice of PEO processing parameters like composition of the electrolyte, process duration, current mode etc., will affect the biological behavior of the bio-medical implant, as they determine the structural and functional characteristics of the coatings.

PEO functionalized Mg–Mn–Ce alloy was investigated by Gnedenkov S.V. (2014) *et al.* (Gnedenkov et al., 2014), [6], for resorbable implant applications. The obtained PEO hydroxyapatite coatings were additionally subjected to a post-treatment by incorporating super dispersed polytetrafluoroethylene (SPTFE) powder in the coating's pores, thereby obtaining multifunctional and corrosion resistant composite surfaces. The aqueous electrolyte was composed of 25 g/L calcium glycerophosphate, 5 g/L sodium fluoride and 20% sodium hydroxide solution for pH adjustment. A 10 min bipolar mode PEO process was employed. The potential for anodic component was raised up to 300V with a rate of 0,45V/s, while the potential for the cathodic mode was set at 30V. Their work demonstrated the possibility to use PEO processes to obtain porous hydroxyapatite coatings with good Ca/P ratio. The corrosion rate of the protective coatings was reduced furthermore by using SPTFE as pore sealant.

Enhanced mechanical properties and increased corrosion resistance of biodegradable PEO coated AZ31 alloy were reported *by White and coworkers (2016)* (White et al., 2016), [7]. A two minutes PEO process, with a pulsed AC power source and a maximum current density of 118 mA/cm^2, was applied to AZ31 alloy samples. Their experimental variable was the electrolyte composition. Thus, they used 3 electrolytes, all of them containing 0.08M sodium silicate and 0.05M sodium hydroxide. Furthermore, the additives of the first electrolyte were 0.1M potassium fluoride and 0.04 sodium dihydrogen phosphate, the second one had no potassium fluoride and the last one had potassium 0.1M fluoride but no sodium dihydrogen phosphate. PEO-samples, with/without phosphate and fluoride, exhibited improved mechanical properties compared to the unprocessed AZ31 alloy. Fluoride containing electrolyte produced denser coatings with lower roughness. The potassium fluoride was also responsible for better corrosion properties.

Yang et al. (2017) [8] used different concentrations of hydroxyapatite particles (HAp) in a phosphate-based electrolyte (10 g/L Na3PO4 and 2 g/L NaOH) to improve the corrosion behavior of pure magnesium. The PEO coatings of Mg substrate were obtained using a DC power source in pulsed mode (20% duty cycle) at a frequency of 200 Hz and current density of 50 mA/cm2, for 10 minutes. Their results showed a reactive incorporation of HA particles in the coatings, forming a magnesium-calcium-phosphate phase. The addition of HA particles improved the corrosion resistance of the PEO coating by inducing the formation of denser layer.

Fu L (2019) [9] studied the corrosion resistance and cytocompatibility of PEO coated AZ31 Mg alloy as a function of incorporated fluorine content. They used a bipolar pulsed current source in galvanostatic mode for 10 minutes processing time. The electrolyte was composed of 3g/L of Sodium Hexametaphosphate, 2g/L sodium hydroxide and different concentrations of potassium fluoride (0, 5, 10, 15 g/L). By investigating the structure, composition, corrosion and cytotoxicity of the as-obtained coatings, they found that the use of an electrolyte containing 10 g/L potassium fluoride, gave the best results regarding the cell compatibility and corrosion resistance.

Corrosion resistance and cytocompatibility as a function of 4 independent PEO parameters (four operating factors include treatment time, duty cycle, current density, and pulsing frequency) was investigated by Jian et al. (2019) [10], for the same AZ31 alloy. They used an aqueous electrolyte composed of 60 g/L Na_2SiO_3, 20 g/L Na_3PO_4, and 70 g/L NaOH. The coatings thus obtained were characterized for surface composition and morphology, electrochemical behavior in Hanks solution and in vivo-testing. In vivo testing showed the improved cytocompatibility and anticorrosive properties of the magnesium silicate containing PEO coatings. The optimal PEO parameters for targeted properties were found to be: current density of 200 mA/cm^2, at a frequency of 4 KHz, duty cycle of 20%, and 30 min process duration.

PEO processed pure Mg was also studied by Y.K. Kim et al. (2019) [11], which used hyaluronic acid (HA) and Carboxymethylcellulose (CMC) to seal the pores and improve bonging properties of the oxide coatings. A unipolar pulsed DC current (galvanostatic mode with 0.3A/cm^2 current density) and an electrolyte composed of 0,1M sodium phosphate, 1M sodium hydroxide and 0,1M glycerol solutions were the main PEO process parameters. Magnesium PEO coated samples were subsequently immersed in a 0.1 mg /ml hyaluronic acid commercial solution for 20h at 900C. Surface characterization, scratch and immersion testing in SBF, cytotoxicity assessment and in-vivo testing on rats, showed that the chemical bonding of the oxide layer with the polymer increased the anti-corrosion properties of immersed PEO coatings, although magnesium and HA were absorbed very quick during the in-vivo testing. They in-vivo testing showed that the thickness and retention of the coatings are crucial for the control of the initial corrosion of magnesium implants.

J. da Silva Rodrigues et al. (2021) [12] studied the anti-corrosion and tribological properties of ZK30 magnesium alloys coated by plasma electrolytic oxidation. The electrolyte used in their experiments was a mixture of 17.5 g/L sodium metasilicate, 4 g/L potassium hydroxide and 3 g/L sodium borate. The samples were subjected to a 60 minutes PEO process, using an alternative current source at a frequency of 1KHz (50% duty cycle) and a maximum 240 V electric potential. The main phases composing the PEO coatings were magnesium silicate and magnesium oxide. Their presented results clearly indicated that the corrosion and wear resistance increased for the PEO coatings compared to uncoated samples.

Application of pure Mg for biodegradable implants was studied by Y. Husak et al. (2021) [13]. The bioactive coatings were obtained using silicate-based electrolytes with the following composition: 10 g/L sodium silicate, 5 g/L sodium fluoride, and 10g/L sodium hydroxide for one electrolyte and, 10g/L calcium hydroxide for another one. A pulsed DC current regime was used for PEO processing of pure Mg surface. After a first galvanostatic stage (100 mA/cm2 current density), a potentiostatic mode was employed for an overall process time of 10 minutes. Their findings showed that the calcium hydroxide was a crucial constituent of the experiment, the PEO process in calcium-based electrolyte producing a higher active surface area with better biocompatibility, but with poor antimicrobial activity.

PEO coatings on AM50 alloy with tuned degradation rate [14] were developed by Lu and coworkers in 2022. In their study, the Mg underwent a DC pulsed PEO process for 20 minutes, at a frequency of 250 Hz, at a fixed current density of 0.37mA/cm^2. The maximum applied electric potential was about 420V. Insoluble additives consisting in different concentrations (5 and 10 g/L) of SiO_2 nanoparticles were added to an electrolyte composed of 20 g/L Na_3PO_4 and 8 g/L KOH. Same electrolyte without nanoparticles was used in the PEO process for the purpose of comparison. They found that the addition of SiO_2 nanoparticles to the electrolyte changed the microstructure and phase composition of the coatings, the crystalline magnesium oxide phase turning to amorphous at increased nanoparticles concentration in the electrolyte. The nanoparticles containing coatings were capable of delaying the initial stage of Mg alloy degradation by a mechanism in two stages: first a dissolution of the unstable amorphous material, followed by a reaction with calcium ions from body fluid and redeposition on the materials surface.

2.2. Plasma electrolytic oxidation of Mg and its alloys for Engineering applications

Although magnesium is the lightest engineering metal, in its pure form can not be used for engineering and structural applications mainly because it corrodes quickly hence the need for alloying and coating. Most common alloying elements used for engineering applications are aluminum and zinc which improve the material's strength, hardness, and slows down its corrosiveness. Most common are AZ31, which has low mass density and good mechanical properties, and AZ91 which has high strength, good castability and corrosion resistance, most suitable for aerospace applications. Also, cast magnesium alloys like Mg-Al-based (AZ, AM) and Mg-rare earth-based (WE43, E21, ZK) are the most used type for automotive applications. Lately, a particular importance is paid to obtaining ceramic and nanostructured coatings on common Mg alloys by PEO with enhanced wear resistance, corrosion protection, thermal control and frictional behavior. For PEO usage on Mg alloys targeting engineering applications most important parameters are: selecting proper Mg alloy, electrolyte composition and PEO process parameters.

Some recent studies involving PEO processing of Mg and its alloys for engineering applications are presented below.

Malayoğlu U and Tekin K C (2015) [15] studied the wear resistance improvement of PEO coated rare earth (RE) containing Mg alloys E21 (AMS 4429) and WE43 (AMS 4427) for aerospace and automotive industries applications. In order to also obtain a lubricant top layer, the surface of PEO coatings was post-treated by spray method with MoS2 lubricant. For PEO coating they used a commercial system with a 25 kW pulsed bipolar alternating current power supply and an alkaline electrolyte composed of $Na_2SiO_3.5H_2O$ (5 g/L), Na3PO4.12H2O (2 g/L), KOH (3 g/L) and KF (2 g/L). The coating treatment was conducted for 20 min in constant current mode with an initial current density of 10 A dm-2. The PEO coating thickness calculated by cross section SEM analysis was 26.6±4.2 and 26.5±3.1 μm respectively while the main phase identified by XRD corresponded to periclase (MgO) and forsterite (Mg2SiO4). Both scratch and wear tests showed that PEO coating greatly improved the wear resistance of the substrate alloys with a further improvement achieved by the adherent spray MoS2 lubricant layer.

Madhan Kumar A et al. (2015) [16] studied the corrosion protection behavior of single and double PEO coating of AZ31 Mg alloy targeting the aerospace applications. PEO treatments were carried out using pulsed DC electrical power source, 7 mA/cm2 current density, 1200 s total oxidation time, 500 Hz pulse frequency with 50% duty ratio. As electrolytes was used aqueous solution of KOH (0.03 M), KF (0.05 M), $Na_3(PO_4)_2$ (0.05 M) and Na_2SiO_3 (0.05 M). For obtaining single and dual PEO coating, they performed the oxidation only in silicate, respectively phosphate electrolytes and then using two stage sequential oxidation. SEM analysis showed highest porosity for PEO coating processed in silicate electrolyte, while XRD analysis indicated MgO and Mg_2SiO_4, respectively $Mg_3(PO_4)_2$ as main crystalline phases present within the obtained coatings. The average micro hardness results of the coated substrates (~300 HV) proved to be 4-5 times higher than that of the substrate, while the corrosion tests showed best performance for the dual layer silicate ⇄ phosphate PEO ceramic coatings.

Zhongping Yao et al. (2016) [17] reported the obtaining of PEO coatings on Mg–Li alloys using phosphate electrolyte for thermal control (high emissivity and low absorptance) with direct applications in aircraft and spacecraft industries. PEO treatments were carried out using a single pulse power generator of 10 kW and phosphate-based electrolyte of 3.75–7.50 g s $(NaPO_3)_6$, 0.75–1.5 g $(NaPO_3)_n$, 3.0– 6.0 g/L NaOH and 0.75–1.5 g/L NaF. Total PEO process time was 30 min, performed at constant current density (i) of 3, 5, 7 and 10 A/dm^{-2}, with 20% duty cycle of pulses and 500 Hz working frequency. XRF phase analysis showed the presence of crystallized $Mg_2(PO_4)_3$ and MgO within the obtained PEO coatings, while SEM-EDS and surface roughness (Ra) tests showed an increase in coating's thickness and roughness proportional with the current density and reaction time, while the increase in electrolyte concentration produce a higher surface roughness but also deceases the coating's thickness. The authors report as optimal conditions for thermal control properties of PEO coating having the lowest solar absorptance (0.33) and the highest infrared emissivity (0.85) to be achieved when using 10 A/dm^2, for 20 min in 8.250 g/L phosphate solution.

Bih-Show Lou et al. (2017) [18] report the usage of Si_3N_4 nanoparticle additives in PEO of AZ31 magnesium alloys in order to obtain coating with improved mechanical, wear and corrosion resistance properties which makes the material more suitable to be used in the aerospace, automotive and personal computer industries. For the PEO process it was used a DC power supply unit with a pulse generator unit which provided a 50% duty cycle and 1000 Hz frequency, a constant current density of 83.3 mA/cm^2 under a unipolar + mode for 10 min and a maximum voltage of 475 V. The aqueous electrolyte solution consisted of 6 g/L K_3PO_4 and 20 g/L NaAlO2 in which they added from 1 to 4 g/L Si_3N_4 nanoparticles with 50 nm average particle size, together with 100 mL/L alcohol dispersant and 1 g/L SDS anionic surfactant. XRD data showed the obtaining of $MgAl_2O_4$ and MgO crystalline phases, with the addition of Mg_2SiO_4 once the Si_3N_4 nanoparticle have been added. The authors report optimum PEO synthesis properties for best corrosion resistance, high hardness, good adhesion and tribological performance to be obtained by adding 2 g/L Si_3N_4 nanoparticle within the electrolyte.

Lorella Ceschini et al. (2017) [19] evaluated the rotating fatigue comportment of EV31A Mg alloy enriched with rare earth (RE) treated by PEO with targeted applications in aerospace and automotive industries. The RE rich magnesium alloy (Nd up to 3.1 wt% and Gd up to 1.7 wt%) have been commercially acquired. Test samples have been obtained by sand casting, machined to specific geometry, heat-treated and finished with emery papers to two different surface roughness of Ra ≈ 0.6–0.8 μm and Ra ≈ 0.2–0.3 μm. The samples have then been subjected to a PEO treatment using a proprietary production process. The PEO obtained samples have then been sandblasted in order to obtain different surface roughness (Ra) between 1.9 and 2.1 μm and then subjected to rotating fatigue tests. Data showed that the residual tensile stresses induced by the PEO treatment causes a 15% decrease in the fatigue strength (88 MPa) comparing to the bare alloy (103 MPa) while the effect of surface roughness was negligible. Also, there are no differences between the untreated and PEO-treated alloy, regarding the mechanisms of crack initiation.

Bih-Show Lou et al. (2018) [20] studied the effect of MoS2 particle addition to the mechanical properties and corrosion resistance capabilities of PEO AZ31 magnesium alloys in order to extend their applicability beyond aerospace, automotive and personal computing industries. For the PEO process it was used a DC power supply with a pulse generating system which provided a current density of 83.3 mA/cm^2 and a maximum voltage of 475 V, 50% duty cycle, a ratio Ton/Toff of 500/500 μs, with a total plasma electrolytic oxidation treatment time of 10min. The aqueous electrolyte solution consisted of 6 g/L K_3PO_4 and 20 g/L $NaAlO_2$ in which different concentration (of 2.5, 5.0, 7.5, and 10.0 g/L) of plate-like MoS_2 nanoparticles with 20 nm average thickness have been added. One PEO sample was obtained without nanoparticles (NPs) addition and used it as reference. XRD data showed a major $MgAl_2O_4$ and a minor MgO crystalline phase for the reference sample, with an enhancement of MgO phase proportional with MoS_2 NPs addition. Both hardness and elastic modulus of coatings increased proportional with MoS_2 NPs addition (from 11.2 to 16.6 respective from 107 GPa to 152 GPa). Although the short-term corrosion resistance of Mg alloy greatly increases only by PEO treatment (without or with NPS addition), best results for short-term corrosion resistance, excellent adhesion and wear resistance, have been reported when using 2.5 g/L MoS_2 NPs within the electrolyte, while adding more than 5 g/L leads to the NPs segregation, thus destroying the oxide layer's mechanical properties. On the other hand, MoS_2 NPs addition deteriorate the porous microstructure and reduces the amount of $MgAl_2O_4$ phase, thus having a negative effect on the long-term anticorrosion performance which proved to be best for the PEO sample of AZ31 magnesium alloys obtained without MoS_2 NPs addition.

Tonelli L et al. (2018) [21] studied the effect of graphite NPs addition to PEO coated rare earth (RE) containing Mg alloys EV31A in order to improve its dry sliding comportment against AISI 52100 steel in air with direct applications in aerospace, military and automotive industries. For the PEO process it was used a DC power supply of 400 V/8 A capacity, constant current of 0.5 A cm^{-2}, and different treatment time of 30, 60 and 90 s. A fluoride-free PEO electrolyte was used containing 50 g/L of $Na_5P_3O_{10}$, 50 g/L of Na2SiO3, 40 g/L of NaOH dispersed in deionized water. 3.5 g/L graphite NPs have been dispersed in acetone and added to the electrolyte, together with 0.5 g/L sodium dodecyl sulphate used as surfactant. XRD data indicated as main crystalline phases within the coating layer to be MgO, $Mg_3(PO_4)_2$ and Mg_2SiO_4. SEM-EDS and microhardness analyzes showed that PEO incorporated graphite NPs lead to a higher coating thickness (from 47 μm for PEO without C NPs up to 62 μm for PEO with C NPs) and microhardness (HV0.1 378 for PEO without C NPs up to 483 for PEO with C NPs) which are also proportional with the increase in PEO process treatment time. Also, the surface roughness increases proportionally with PEO treatment time (Ra from 3.8 to 7.5 μm), although graphite NPs addition did not play a significant part in surface roughness but contribute instead to the decreasing of friction and wear. Tribological test showed best results for optimal thickness, compactness, roughness and microhardness for the PEO coatings with graphite NPs addition processed for 60s.

Buling A and Zerrer J (2019) [22] investigated the corrosion and wear behaviors of ultraceramic coatings on AM50, AZ31 and E Form (AZ31 with additional Ca) Mg alloys obtained by PEO using asymmetric bipolar pulsed waveform targeting the applications within the automotive industry. For the PEO process it was used a pulsed DC power source which generated two waveforms with asymmetric bipolar pulse with a controlled gradual slow increase in the anodic branch (82% duty cycle), holding the potential, and an abrupt decrease in the cathodic branch (18% negative cycle) - original waveform which have been applied for PEO of AM50, AZ31, and an optimized waveform with additional on-top signal which have been applied for PEO of E Form Mg alloy. The PEO process parameters used were 290 V for the RMS value of the potential, 500 Hz pulse frequency and different processing time of 3,5 and 9 min. As electrolyte it was used an additivated low-alkaline aqueous solution of 4 g/L NaOH, 15 g/L Na_2SiO_3 and 15 g/L Na4P2O7. Test results show optimal corrosion and wear protection for a coating thickness of ≈25μm for all three Mg alloys PEO processed by the original process while for the optimized PEO ultraceramic process this protection can be achieved at thinner coating thickness of ≈10μm due to a smoother coating with smaller surface pore size.

Morri A et al. (2020) [23] studied the fatigue behavior of PEO magnesium–zinc–zirconium (ZK60A-T5) alloy used in motorsport and aerospace industries due to their high specific strength. The testing specimens had been cut from racing motorbikes forged wheels, subjected to PEO treatments and powder coated by electrostatic painting for corrosion protection and aesthetics. Although the authors do not provide much information about the PEO process parameters, electrolytes (other than alkaline solution was used, containing P and Zr-based compounds) and phase analysis of the obtained PEO coating, they present optical (OM) and scanning electron microscopy (SEM) data. They obtained typical PEO coatings morphologies with pores, microcracks, and discharge microchannels while the rotating bending fatigue tests showed a PEO-induced fatigue strength reduction of 14 to 17% at 107 cycles comparing to the untreated ZK60A-T5 Mg alloy.

Kun Qian et al. (2020) [24] extended the research of PEO coated AZ91D Mg alloy by studying the effect of phosphate-based sealing post-treatment on the material's corrosion performance with a wide range of possible applications within the aerospace, automotive, communication and biomedicine industries. For the PEO process it was used a pulse reverse power which generated a frequency of 500 Hz and a duty ratio of 30 at a constant current of 2 A/dm^2 for a 15 min process tine. As electrolyte was used an alkaline solution of 10–20 g/L Na_2SiO_3, 1–4 g/L KF, and 1–5 g/L NaOH. The PEO coated samples were then sealed in a solution of NaH2PO4 with the addition of sodium fumarate for 15, 30 and 60 min. XRD phase analysis on PEO coated AZ91D alloy shows MgO and Mg_2SiO_4 crystalline phases, with the addition of MgHPO4 discovered after the sealing post-treatment. Corrosion tests show improved anti-corrosive behavior of PEO coated AZ91D alloy after applying the sealing post-treatments with corrosion inhibitor whit best results for the sample which have been sealed for 15 min.

Dmitry V. Mashtalyar et al. (2020) [25] report the obtaining of hard wearproof PEO coatings on MA8 magnesium alloy (1.5 wt.% Mn; 0.15 wt.% Ce; Mg – balance) by TiN NPs addition which make possible to expand the possible applications in the fields of aviation and automotive industries. PEO process was carried out using a conventional reversible thyristor rectifier as power supply generating pulses with 300 Hz frequency and 50% duty cycle with a PEO process in a two-stage bipolar mode. As electrolyte was used sodium fluoride (5 g/L) and silicate (15 g/L) with the addition of TiN nanopowders having 20 nm particle size in different concentrations of 1, 2, 3, and 4 g/L. Data show the presence of TiN, TiOxNy, and TiO2 within the PEO coatings, with TiN NPs being distributed along the PEO-coating and filling its pores. The highest protective properties proved to be for a TiN NPs concentration of 2–3 g/L while the microhardness and wearproof properties of the TiN NPs containing PEO coating doubled comparing to the PEO coatings obtained without TiN NPs.

Xinyan Wang et al. (2021) [26] studied the influence on the corrosion resistance and thermal control property of nano-sized ZnO particles addition in PEO of AZ91 Mg alloy with a focused application in the field of spacecraft and electronics. For PEO process was used DC pulsed power supply generating a 30 mA/cm^2 constant current, 300 Hz pulse frequency and 30% duty ratio, with different treatment time of 15, 25 and 35 min. The electrolyte consisted of a silicate solution (10 g/L Na_2SiO_3, 5 g/L KF and 2 g/L NaOH) and a phosphate solution (10 g/L Na_3PO_4, 5 g/ L KF and 2 g/L NaOH) with the addition of 10 g/L nano-sized ZnO particle with average particle size of 50 nm. XRD phase analysis showed the presence of MgO, Mg_2SiO4 and ZnO for PEO in silicate electrolyte and ZnO, MgO and $Mg_3(PO_4)_2$ for PEO in phosphate electrolyte [27,28]. Data also show ZnO NPS being inertly incorporated into the PEO coating which results in low porosity surfaces and enhanced corrosion resistance, also having a major impact on coating's microstructure, morphology and phase composition [29]. The ZnO NPs addition in PEO of AZ91 Mg alloy induces a decrease in coating's absorptance and an increase in emissivity.

3. Conclusions

Mg alloys have wide applications in various industries, especially medicine and engineering, due to their lightweight design, which can greatly improve the efficiency of bioresorbability, biocompatibility, reduce fuel consumption and reduce vehicle emissions. However, the high sensitivity of Mg alloys to corrosive media limits their use. Surface functionalization by PEO can become one of the most effective and economical ways to increase properties of Mg alloys. The paper highlights the advantages of the PEO technique as surface treatment process on Mg alloys to produce thick and dense metal oxide coatings, especially on light metals, primarily to improve their wear and corrosion resistance. Several investigations presented in the principal database have been carried out to study the influence of the process parameters on coating performance developed through the PEO process on Mg alloys for medial and engineering applications.

References

[1] Yang Y, Xiong X, Chen J, Peng X, Chen D and Pan F 2021 *Journal of Magnesium and Alloys* **9** 705–747
[2] Tan J and Ramakrishna S 2021 *Applied Sciences (Switzerland)* **11**
[3] Li N, Zheng Y 2013 *Journal of Materials Science and Technology* **29** 489–502
[4] Echeverry-Rendon M, Allain J P, Robledo S M, Echeverria F and Harmsen M C 2019 *Materials Science and Engineering C* **102** 150–163
[5] Prasadh S, Ratheesh V, Manakari V, Parande G, Gupta M and Wong R 2019 *Metals (Basel)* **9**
[6] Gnedenkov S V, Sinebryukhov S L, Zavidnaya A G, Egorkin V S, v Puz' A, Mashtalyar D V, Sergienko V I, Yerokhin A L and Matthews A 2014 *J Taiwan Inst Chem Eng* **45** 3104–3109
[7] White L, Koo Y, Neralla S, Sankar J, Yun Y 2016 *Materials Science and Engineering B: Solid-State Materials for Advanced Technology* **208** 39–46
[8] Yang J, Lu X, Blawert C, Di S and Zheludkevich M L 2017 *Surface and Coatings Technology* **319** 359–369
[9] Fu L, Yang Y, Zhang L, Wu Y, Liang J and Cao B 2019 *Coatings* **9**
[10] Jian S Y, Ho M L, Shih B C, Wang Y J, Weng L W, Wang M W and Tseng C C 2019 *Coatings* **9**
[11] Kim Y K, Jang Y S, Kim S Y and Lee M H 2019 *Applied Surface Science* **473** 31–39
[12] da Silva Rodrigues J, Marasca Antonini L, da Cunha Bastos AA, Zhou J and de Fraga Malfatti C 2021 *Surface and Coatings Technology* **410**
[13] Husak Y, Michalska J, Oleshko O, Korniienko V, Grundsteins K, Dryhval B, Altundal S, Mishchenko O, Viter R, Pogorielov M and Simka W 2021 *Molecules* **26**
[14] Lu X, Blawert C, Luthringer B J C and Zheludkevich M L 2022 *Frontiers in Chemical Engineering* **4**
[15] Malayoǐlu U and Tekin K C 2015 *Surface Engineering* **31** 526–533
[16] Madhan Kumar A, Kwon S H, Jung H C and Shin K S 2015 *Materials Chemistry and Physics* **149** 480–486
[17] Yao Z, Ju P, Xia Q, Wang J, Su P, Wei H, Li D and Jiang Z 2016 *Surface and Coatings Technology* **307** 1236–1240
[18] Lou B S, Lin Y Y, Tseng C M, Lu Y C, Duh J G and Lee J W 2017 *Surface and Coatings Technology* **332** 358–367
[19] Ceschini L, Morri A, Angelini V and Messieri S 2017 *Metals (Basel)* **7** 1–14
[20] Lou B S, Lee J W, Tseng C M, Lin Y Y and Yen C A 2018 *Surface and Coatings Technology*

350 813–822
[21] Tonelli L, Pezzato L, Dolcet P, Dabalà M and Martini C 2018 *Wear* **404–405** 122–132
[22] Buling A and Zerrer J 2019 *Surface and Coatings Technology* **369** 142–155
[23] Morri A, Ceschini L, Martini C and Bernardi A 2020 *Coatings* **10** 1–13
[24] Qian K, Li W, Lu X, Han X, Jin Y, Zhang T and Wang F 2020 *Journal of Magnesium and Alloys* **8** 1328–1340
[25] Mashtalyar D V, Sinebryukhov S L, Imshinetskiy I M, Gnedenkov A S, Nadaraia K V, Ustinov A and Gnedenkov S V 2020 *Applied Surface Science* **503** 144062
[26] Wang X, Lu X, Ju P, Chen Y, Zhang T and Wang F 2021 *Surface and Coatings Technology* **409** 126905
[27] Ciubară A, Burlea ŞL, Axinte M, Cimpoeşu R, Chicet DL, Manole V, Burlea G and Cimpoeşu N 2018 *IOP Conference Series: Materials Science and Engineering* **374(1)** 012066.
[28] Nergis D P B, Cimpoesu N, Vizureanu P, Baciu C and Bejinariu C 2019 *Materials Today: Proceedings* **19** 969–978
[29] Baltatu MS, Vizureanu P, Cimpoesu R, Abdullah MMA and Sandu AV 2016 *Revista de chimie* **67(10)** 2100-2102.

Influence of heat treatment on the mechanical properties of alkali-activated fly ash-based binders with marble dust substitution

B A Ionescu[1,2], C Mircea[3], M Chira[1] and A-V Lăzărescu[1,2]

[1]NIRD URBAN-INCERC Cluj-Napoca Branch, 117 Calea Floresti, Cluj-Napoca, 400524, Romania
[2]Technical University of Cluj-Napoca, IOSUD-UTCN Doctoral School, 15 Constantin Daicoviciu Street, Cluj-Napoca, 400020, Romania
[3]Civil Engineering Faculty, Technical University of Cluj-Napoca, 28 Memorandumului Street, 400114 Cluj-Napoca, Romania

E-mail: bradut.ionescu@incerc-cluj.ro

Abstract. Marble waste contains a high level of calcium, which is obtained from the cutting process in marble production. The properties of geopolymer binders are influenced not only by the amount of alkali activators, their ratio, the molarity used, and the Si and Al content of the mineral additives used in the mixture, but also by the duration of the heat treatment and the heat treatment temperature. This study aimed to produce alkali-activated geopolymer binders based on fly ash and marble dust. Alkali-activated geopolymer pastes were made both with heat treatment for 24h at 70°C and without heat treatment at (23±2) °C. Structural analysis by optical microscopy revealed, in the case of heat-treated samples, the formation of pores with an average size of 1÷1.5 mm, much larger than in the case of samples without heat treatment.

1. Introduction

Geopolymerization is an exothermic reaction between aluminum- and silicon-rich precursors of industrial or geological origin with concentrated alkali activators in a combined alkaline silicate and hydroxide solution [1-3], resulting in a three-dimensional polymer chain and ring structure formed by Si-O-Al-O bonds. Geopolymers can be made at a temperature ranging from room temperature or even room temperature up to 100 °C and atmospheric pressure [4-6]. Geopolymers have a wide range of compositions, but of the most popular is Al_2O_3-M_2O-nSiO_2-10 H_2O, where M = Na, K or Cs, and n ranges from 2 to 4 [7].

Geopolymerization is a relatively new technology that converts aluminosilicate materials into geopolymers, which involves a chemical reaction between solid aluminosilicate oxides and an alkaline activating solution at ambient or slightly elevated temperatures, producing an amorphous to semi-crystalline polymeric structure with Si-O-Al and Si-O-Si bonds [8-12]. The geopolymerization process is presented in Figure 1 [13].

Due to alkaline activators, two simultaneous processes take place, namely dissolution into monomeric Al and Si species and polymerisation. OH groups play a main role in the dissolution of Si and Al species, and the presence of alkali ions play a main role in the polycondensation of these species in the polymeric 3D network. Another function of the alkali activator is to maintain the neutral AlO_4^- anions. The most common materials used as activator solutions are Na and K [14,15].

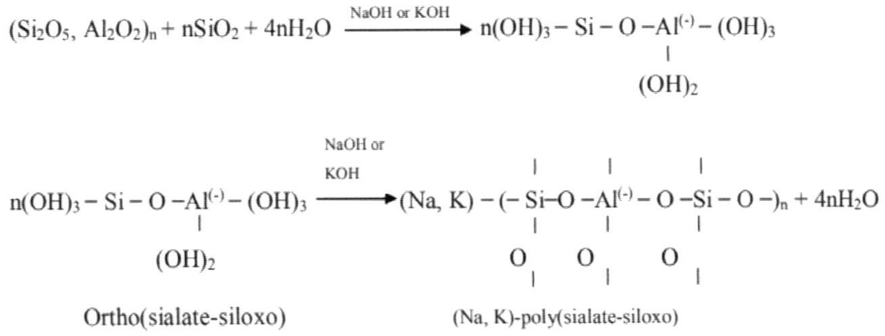

Figure 1. Dissolution of Aluminosilicates, hydroxide (Na, K), water and silicate addition in gel form, followed by dihydroxylation process to form a geopolymer network [13].

According to the literature, NaOH and KOH solutions are the most commonly used to produce the geopolymerisation process. By adding a high pH alkaline solution to a precursor, hidrolysis of Si^{4+} and Al^{3+} components takes place, which will form monomeric species such as $[SiO(OH)_3]^-$, $[SiO_2(OH)_2]^{2-}$ and $[Al(OH)_4]^-$ [16]. The following equations (1), (2), (3) describe the dissolution and hydrolysis of Si and Al species [17]:

$$Al_2O_3 + 3H_2O + 2OH^- \rightarrow 2[Al(OH)_4]^- \quad (1)$$

$$SiO_2 + H_2O + OH^- \rightarrow [SiO(OH)_3]^- \quad (2)$$

$$SiO_2 + 2OH^- \rightarrow [SiO_2(OH)_2]^{2-} \quad (3)$$

Geopolymer manufacturing offers offer a cost reduction of between 10% and 30% compared to the cost of conventional construction technologies. [18]. Marble waste contains a high level of calcium, which is obtained from the cutting process in marble production [19]. The waste resulting from the marble industry is in the form of solid dust or semi-solid sludge, causing storage and environmental problems. Marble dust waste comes either from marble processing or from the excavation of marble deposits. Careless dumping of waste on the land surface without taking precautions causes major environmental problems, especially during dry periods when marble dust dries out and is carried by currents through the atmosphere and then settles on vegetation, causing problems for agricultures crops as well as plants and three. According to the literature, there are several ways to reuse marble waste for certain purposes, such as in the brick industry [20], in transport infrastructure [21], as a material in mixtures with cement [22-25], geopolymer concrete [26] and aggregates or mineral admixtures [27].

Lee et at. [28] studied the possibility of producing geopolymer concrete based on marble and ground-granulated blast-furnace slag (GGBFS) and examined its applicability as a durable structural material in the construction industry. Based on the results obtained from the research it could be demonstrated that the marble-based geopoliymer concrete had very good physical-mechanical properties. It was also found that the marble based geopolymer concrete showed good workability an its properties were not influenced by temperature changes. It can be said that marble-based geopolymer concrete has a high potential for future engineering development.

Following their research and based on the results obtain by Thakur et al. [29] they stated that geopolymer materials made by increasing the molarity of NaOH solutions from 2M to 4M resulted in an increase in compressive strength from 4.61 MPa to 6,52 MPa.Due to the reaction that occurred between the silicon dioxide present in fly ash with the calcium hydroxide in the marble waste lead to the formation of a bond of calcium silicate hydrates that could improve the bonds in geopolymer matrix, leading to an increase in bulk density and also in compressive strength. Based on these results, the realisation of hybrid geopolymers using marble waste represents an economical and sustainable solution for the management of marble waste currently generated in large quantities in different countries of the world.

Osman [30] studied the effect of marble dust on the properties of a geopolymer paste. The use of marble dust, fly ash and silica fume in various different percentages with alkali activator were used to study the mechanical properties (compressive strength, flexural strength and tensile strength) and durability (water absorption) of geopolymer paste. The results showed that the best compressive strength (23.78 N/mm^2) was obtained for a recipe with 70% fly ash and 30% marble dust and w/b=28%. In addition, it was observed that the use of a w/b=0.28 ratio in the formation of geopolymer paste containing 90% fly ash and 10% marble dust gave the highest flexural strength (2.45 N/mm^2). The results were also analysed by SEM imaging (Figure 2).

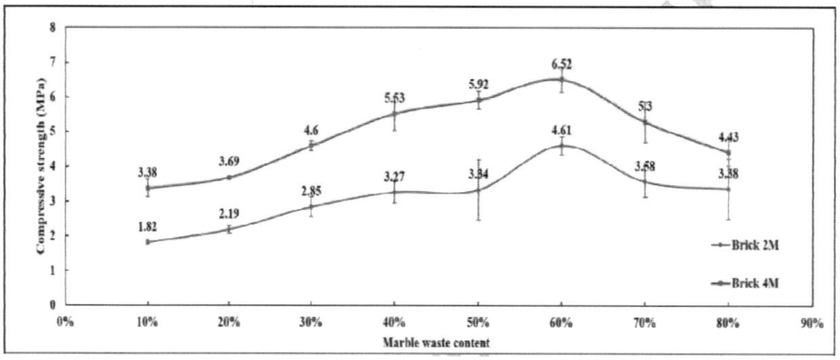

Figure 2. Compressive strength of hybrid geopolymer composite samples prepared with different molarities and marble waste content [29].

In another study, Tekin [31] made a geopolymer paste based on waste marble, volcanic tuff and travertine, alkaline activated with NaOH-solutions. He used different molarity of NaOH (1M, 5M and 10M) opting for a heat treatment with a duration of 24h and temperatures of (20, 45 and 75°C). The highest compressive strength obtaind by him was 46MPa using a molarity of 10M, having 80% BSW (Bayburt stone waste) and 20% TW (travertine waste), the amount of marble waste used being unclear. Okoye et al. [32] found that compressive strength increases up to a certain temperature (100°C), after which it decreases.

Geopolymers have have been found to exhibit better properties when subjected to thermal (accelerated uring, due faster geopolymerisation and polycondensation reaction [33]. Heah et al. in their studies stated that curing of metakaolin-based geopolymers at room temperature was not feasible, while increasing temperature (40, 60, 80 and 100°C) favoured strength increase after 1-3 days.

Rovnanik [34] reported that curing of metakaolin-based geopolymer at elevated temperature (40-80°C) resulted in increase strength, but at the age of 28 days, the mechanical properties were weaker compared to the results obtained for ambient or slightly low temperature.

Vijai et al. [35] studied the properties of geopolymer concrete such as density and compressive strength for two different types of curing; heat curing at a temperature of 60°C for 24 hours in a hot air oven and a curing at an ambient temperature. The source material used in the study was fly ash and a combination of NaOH and sodium silicate were used as chemical activators. However, the compressive strength at the age of 28 days for the two types of geopolymers with and without heat treatment did not vary substantially (23.4 and 33.2 MPa, respectively), showing that the long-term mechanical properties are not affected by the curing regime, as is the case at an early age. In another study Neupane et al. [36] studied the effect of temperature on the compressive strength of geopolymers. Two types of blended geopolymer pastes with fly ash: slag ratios of 7:3 and 4:6 were used to produce geopolymer concrete mixtures. The compressive strength results for both types of fly ash:slag geopolymers (7:3 and 4:6 ratio) at 1 day age were 58MPa and 65.5 MPa for those with heat treatment, respectively 7MPa and 12.5 MPa for geopolymers with room temperature curing.

2. Materials and methods

2.1. Materials

Fly ash used in the production of geopolymer binder was obtained from the Rovinari Thermal Power Plant, Romania. Marble dust was obtained from the cutting and processing of marble slabs. Sodium silicate and NaOH in the form of pearls with a purity of 99.7% were purchased commercially. The chemical composition of fly ash and marble dust is presented in Table 1 while the images of the used materials are presented in Figure 3.

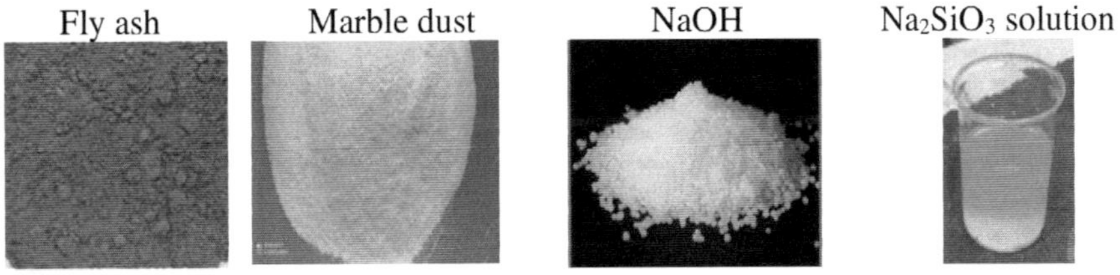

Figure 3. Raw materials used in the production of alkali-activated geopolymer paste.

Table 1. Fly ash and marble dust chemical composition.

Oxides	SiO_2	Al_2O_3	Fe_2O_3	CaO	MgO	SO_3	Na_2O	K_2O	TiO_2	L.O.I
F.A %	46.9	23.8	10.1	10.7	2.7	0.5	0.6	1.7	0.9	2.1
M.D. %	0.28	1.37	0.17	54.63	0.43	-	-	-	-	-

2.2. Synthesis of geopolymer

Geopolymer pastes were made from 90% fly ash and 10% marble dust and an alkaline activating solution by combining NaOH solution (8M) and Na_2SiO_3 solution (Table 2). The Na_2SiO_3/NaOH solution ratio used was 2.5. The procedure of producing the alkali-activated fly ash-marble dust based geopolymer binders are shown in Figure 4. Geopolymer paste samples were both to heat treatment (70°C, 24h) and without heat treatment (23°C). Mixing of fly ash with marble dust was carried out for about 60s for complete homogenization using a paddle mixer. Subsequently, after mixing the fly ash with the marble dust, the alkaline activator solution was poured in gradually over 70s, at first at low speed. Pouring the alkaline activator too abruptly can lead to the instant hardening effect (rapid - abrupt hardening of the geopolymer paste). Mixing of alkaline activators with fly ash and marble dust was done for about 2.5-3 min.

2.3. Testing methods

For each set of specimens with and without heat treatment, the apparent density was determined as the ratio of mass to volume. Bending strength was determined using the three-point bending test (3PB) in accordance with EN 196-1. A digital flexural strength tester suitable for loads up to 10kN (±10%) with a loading rate of (50±10) N/s was used. The testing machine is provided with a bending device consisting of two steel support rollers with a diameter of (10±0.5) mm, arranged at a distance of (100±0.5) mm from each other and a third, centrally placed load roller.

Table 2. Alkali-activated geopolymer pastes mix-design.

Mixture	Fly ash (%)	Marble dust (%)	$Na_2SiO_3/NaOH$	Heat (°C)
R9 **R53FTT**	90	10	2.5	70°C, 24H 23±2°C

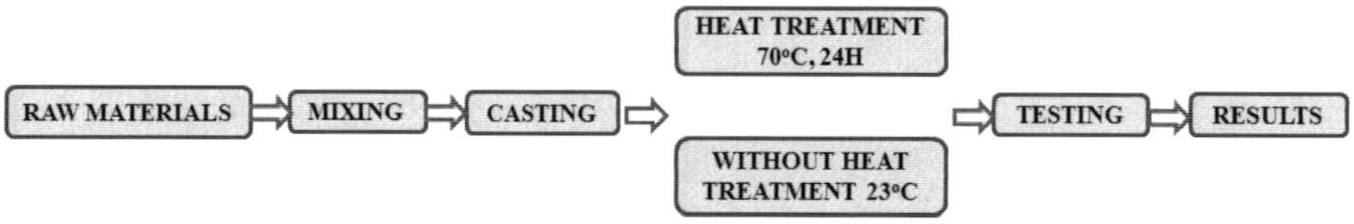

Figure 4. Geopolymer paste production process.

The compressive strength of the specimens was determined in accordance with the provisions of EN 196-1, using the compression test of the prismatic specimen halves resulting from the three-point bending test (3PB). The compressive loading rate used is 50 N/s (0.12 MPa/s). The test age of the specimens was 7 days.

After reaching the age of 7 days after casting, all sets of specimens were tested for flexural strength and compressive strength and subjected to Fourier Transform Infrared (FTIR) analysis for further microstructural analysis. Laboratory tests by the FTIR method were carried out using a Jasco FT/IR-6200 Fourier Transform Infrared Spectrometer on pellets made of solid mixture in the ratio KBr: prepared material = 150:1, pressed under the action of vacuum at a pressure of 5 bar.

Fourier Transform Infrared (FTIR) analysis is widely used to perform analysis of geopolymer pastes. Peaks occurring at different wavenumbers with different intensities explain the type of formations, bonds and vibrations in the material matrix. This helps to identify the different compositions of a material. FTIR analysis was carried out to observe the changes that occur as a function of temperature.

The microstructural study of geopolymer pastes was carried out using a Leica DNC2900 Stereomicroscope.

3. Results and discussions

3.1. Apparent density

It can be seen that the bulk density (Figure 5) at 7 days was 1540 kg/m^3 for mixture R9 with heat treatment and 1620 kg/m^3 for mixture R53FTT without heat treatment. The loss of bulk density after heat treatment and further up to the test age is attributed to the evaporation of water from the mixture, as research shows that, unlike OPC binders, in the geopolymer structure the water in the alkali activator only helps to dissolve the ash and generates the workability of the binder.

3.2. Flexural strength

The bending strength at 7 days of age was carried out on specimens with and without heat treatment with dimensions of 40 x 40 x 160 mm. The bending strength obtained as an average of three geopolymer paste specimens with heat treatment was 3.31 N/mm^2 and without heat treatment was 2.49 N/mm^2 (Figure 6). It is observed that, R9 geopolymer paste with heat treatment (70°C, 24h) have higher bending strength than those without heat treatment (23±2°C).

3.3. Compressive strength

According to the literature, several attempts have been made to study the effect of different curing conditions on the properties of geopolymer pastes. The compressive strength results showed that heat curing greatly affects the early strength of fly ash-based geopolymer, as the compressive strength at 7 days of samples made with heat treatment was about 2.5 times higher than that of samples met at room temperature without heat treatment.

After the bending tensile tests, compressive strength tests were carried out on the geopolymer paste specimen halves on 40 x 40 mm cubes. The compressive strength at 7 days was 28.80 N/mm^2 for mixture R9 (with heat treatment) and 23.40 N/mm^2 for mixture R53FTT (Figure 7). It is possible that, due to the higher temperature (70°C) and the duration of the heat treatment (24h), the bonds of the compounds formed in the geopolymer paste are stronger than in the geopolymer paste without heat treatment (23±2)°C. An influence on the compressive strength is primarily due to the formation of the geopolymer gel, which develops faster in the presence of high temperature (between 30-90°C).

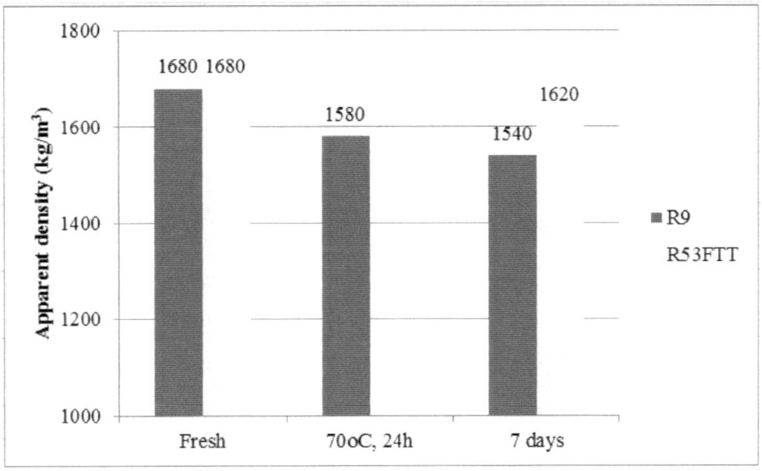

Figure 5. Apparent density of the alkali-activated geopolymer samples.

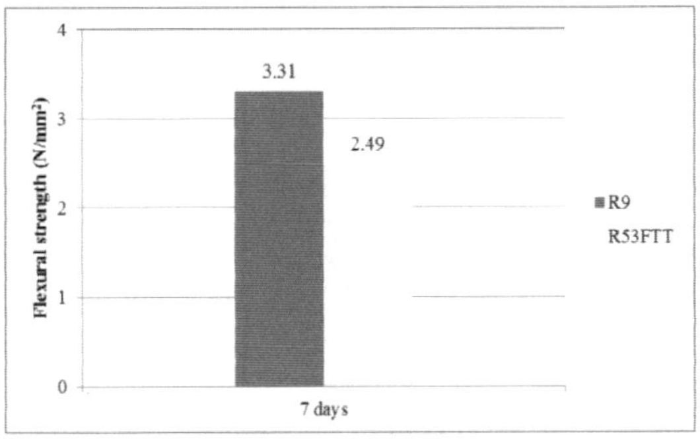

Figure 6. Flexural strength of the alkali-activated geopolymer samples.

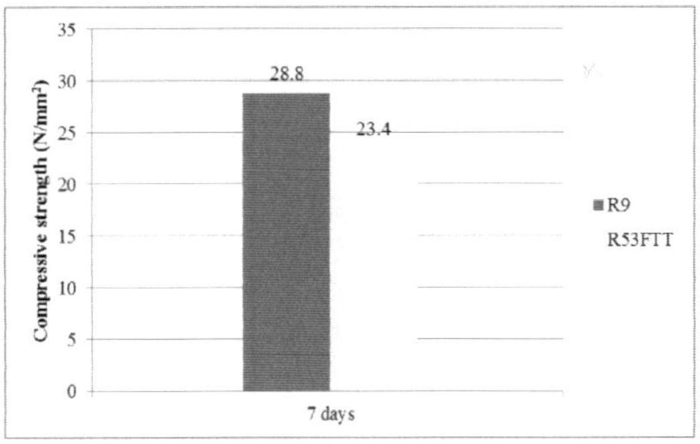

Figure 7. Compressive strength of the alkali-activated geopolymer samples.

3.4. FTIR analysis

FTIR spectra of R9 and R53FTT geopolymer pastes with and without heat treatment are shown in Figure 6 and also reported to other tests in the literature (Table 3).

The intensity of the IR bands decreases over the range 500÷2900 cm^{-1} to 2930÷4000 cm^{-1} as the geopolymer paste making temperature decreases from 70°C, 24h to 23±2°C (Figure 8).

The increased intensity IR bands centered at 1022 cm^{-1} (R9 recipe performed with heat treatment) and 1014 cm^{-1} respectively correspond to deformation vibrations of Si-O-Si or Si-O-Al angles in the alumino-silicate gel and are responsible for the formation of amorphous aluminosilicate gel in binary systems [45,47]. The shift from 1022 cm^{-1} to a lower wavenumber 1014 cm^{-1} corresponds to the Si-

O/Al-O bond formation of the aluminosilicate lattice characteristic of an aluminosilicate gel in amorphous binary systems [51-53]. These bonds are weaker in the sample without heat treatment. The broadened absorption band around 1020 cm^{-1} may be due to the transformation of asymmetric stretching vibrations of Si-O-Si bonds into symmetric stretching vibrations of T-O-Si bonds (T = Si or Al) and into symmetric bending vibrations of T-O-Si bonds (T = Si or Al) and into symmetric bending vibrations of Si-O-Si and O-Si-O bonds in the geopolymer matrix [54-56].

Table 3. Assignment of IR bands in the geopolymer structure according to literature.

Bond	Wave number (cm^{-1})	Reference
Symmetric and antisymmetric O-H bond elongation	~3500 and ~1600	[37, 38]
vibrations in water	3445	[39]
H-O-H deformation vibrations in water	3400 - 3650	[40]
Antisymmetric O-H bond elongation vibrations in water	1650	[41]
	1640	[42]
Elongation vibrations of the C-O bond in the carbonate ion	~1440-1453	[37,43]
Antisymmetric elongation vibrations of the Al-O bond	~1400	[44]
Vibration deformation of Si-O-T angles (T=Si or Al in alumino silicate gel)	1025	[45]
T-O-Si symmetrical elongation vibrations (T=Si or Al)	1020	[46]
Si-O/Al-O of aluminosilicate lattice reflecting amorphous aluminosilicate gel formation in binary systems	1015	[47]
T-O-Si symmetrical elongation vibrations (T=Si or Al)	946	[48, 49]
Si-rich Si-O-Si or O-Si-O deformation vibrations	<461-465	[50]
O-Si-O deformation vibrations	454	[37]

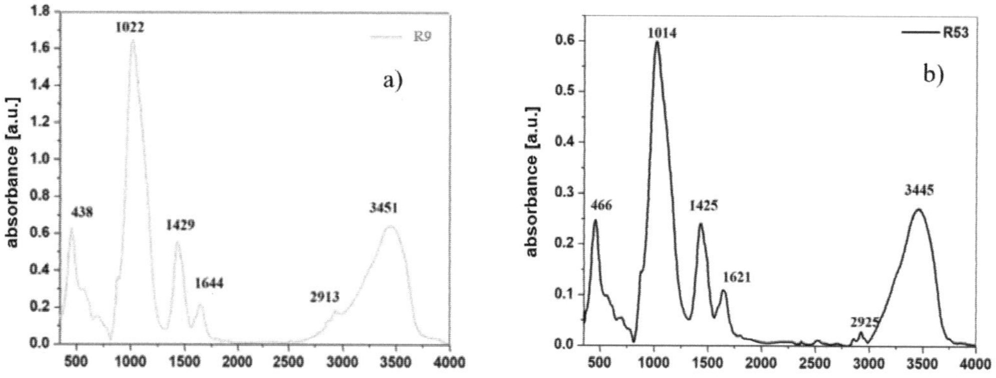

Figure 8. FTIR spectra of geopolymer pastes made with a) heat treatment, b) without heat treatment

The IR bands centered at 1425 cm-1 and 1429 cm-1 correspond to some C-O bond elongation vibrations in the carbonate ion, which are higher in the sample without heat treatment. For sample R53FTT the intensity of the band centered at 1425 cm^{-1} decreases compared to sample R9, indicating the decrease in the number of Al-O bonds.

The asymmetric IR bands 3451 and 3445 cm^{-1}, with shifting tendencies towards higher wavenumbers, correspond to H-O-H deformation vibrations in water [39], vibrations that exhibit antisymmetric elongation of the O-H bond in water [40]. The O-Si-O strain vibrations are higher for the R9 sample without heat treatment, increasing from 438 to 466 cm^{-1}, corresponding to Si-rich O-Si-O, Si-O-Si strain bands.

Following treatment with curing temperatures of 23°C and 70°C, it can be observed that the absorption of peaks showing the existence of aluminosilicates (Si-O-Al) increases with increasing curing temperature. The increase of this absorbance representing higher conversion of aluminosilicate formation is consistent with the increase of the compressive strength of the geopolymer samples.

3.5. Microscopic analysis

In order to characterize the microstructure of the alkaline activated geopolymer pastes and to determine the contribution of temperature to the formed products, R9 geopolymer paste made with heat treatment was compared to R53FTT geopolymer paste made without heat treatment. The analysis was performed using a Leica DNC2900 Stereomicroscope at 1x resolution.

Analysing the pictures obtained by optical microscopy Figure 9, and corroborated with the evolution of the apparent density, from the casting to the age of 7 days, it is estimated that a large amount of water in the geopolymer paste subjected to heat treatment has evaporated, leading to the formation of pores of approx. 1-1,5mm.

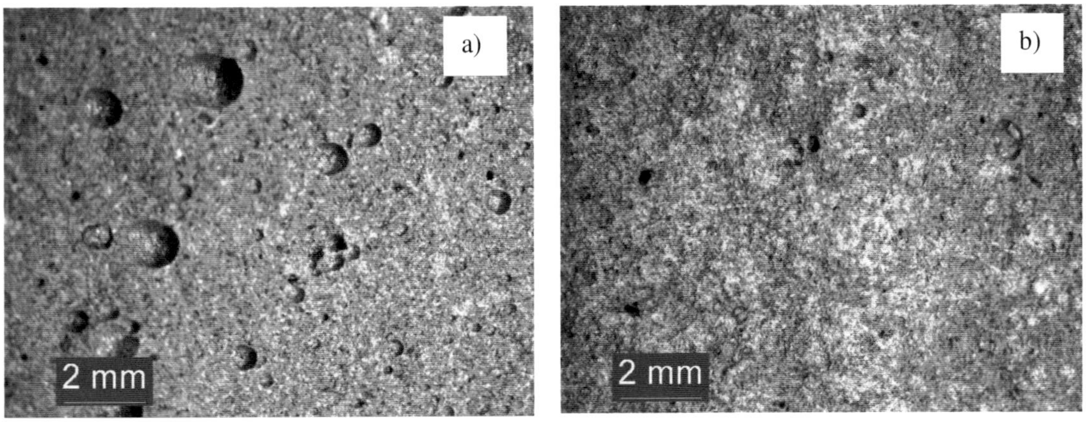

Figure 9. Microscopic analysis of samples a) R9 mixture with heat treatment and b) R53FTT mixture without heat treatment at 1x resolution.

Comparing the two microscopic pictures it can be seen that in the case of heat-treated sample R9 (70°C, 24H), the size and number of pores is larger than in the case of sample R53FTT without heat treatment which is in correlation with the lower bulk densities at 7 days of age. It is estimated that the heat treatment facilitates the formation of the geopolymer gel through a more uniform and homogeneous process, hence a better compressive strength, but influences the structural microscopy.

4. Conclusions

In this work, a physical-mechanical (bending and compression) and microstructural analysis was carried out using a stereomicroscope in order to better understand the effect of temperature on the production of geopolymer pastes made with heat treatment (70°C, 24h) or at room temperature (23±2°C).

The flexural and compressive strength was higher for the heat-treated sample R9, which is in line with the literature.

The compressive strength increases from 23.4 MPa to 28.8 MPa with increasing curing temperature from 23°C to 70°C. The increase in compressive strength is consistent with FTIR analysis, where the IR band increases from 1014 to 1022 cm^{-1}. It was observed that the absorption of peaks showing aluminosilicate (Si-O-Al) increases with increasing curing temperature.

Microscopic analysis revealed large pores in the heat-treated sample, most likely formed by more intense and rapid evaporation of water at 70°C.

Acknowledgments

This research was financially supported by the Project "Entrepreneurial competences and excellence research in doctoral and postdoctoral programs - ANTREDOC", project co-funded by the European Social Fund financing agreement no. 56437/24.07.2019.

Partial support was received from Programme Research for sustainable and ecological integrated solutions for space development and safety of the built environment, with advanced potential for open innovation – "ECOSMARTCONS", Programme code: PN 19 33 04 02: "Sustainable solutions for ensuring the population health and safety within the concept of open innovation and environmental preservation" and Programme code PN 19 33 03 01: "Researches to achieve the acoustic and thermal comfort inside the buildings, using an innovative tool for choosing the optimum structures of construction elements, from classical versus modern materials", financed by the Romanian Government.

References

[1] Rashad A M, and Zeedan S R 2011 *Constr. Build. Mater.* **25** 3098–3107
[2] Palacios M, and Puertas F 2007 *Cem. Concr. Res.* **37** 691–702
[3] Kuenzel C, Ranjbar N 2019 *Resour. Conserv. Recycl.* **150** 104421
[4] Cioffi R, Maffucci L, and Santoro L 2003 *Resour. Conserv. Recycl.* **40** 27–38
[5] Panias D, Giannopoulou I P, and Perraki T 2007 *Colloid Surf. A Physicochem. Eng. Asp.* **301** 246–254
[6] Komnitsas K, and Zaharaki D 2007, *Miner. Eng.*, **20** 1261–1277
[7] Kriven W M, Bell J, Gordon M and Wen G 2005 *Geopolymer, Green Chemistry and Sustainable Development Solutions* 179–183
[8] Davidovits J and Orlinski J 1988 *Proceedings of the 1st International Conference on Geopolymer '88* Compiegne, France **1** 19–24
[9] Davidovits J 1991 *Journal of Thermal Analysis* **37**(8) 1633–1656
[10] Davidovits J 1994a *First International Conference on Alkaline Cements and Concretes,* Kiev, Ukraine 131–149
[11] Davidovits J 1999 *Geopolymer '99 International Conference*, France, 9–40
[12] Duxson P, Mallicoat S W, Lukey G C, Kriven W M and Van Deventer J S J 2007 *Colloids and Surfaces a-Physicochemical and Engineering Aspects* **292** 8–20
[13] Wallah SE, and Rangan BV 2006 *Research Report GC 2* **1**(1) 5
[14] Rodriguez E, Mejia De Gutierrez R and Bernal S 2009 *Revista Facultad De Ingenieria-Universidad De Antioquia* 30–41
[15] Provis J L, Yong C Z, Duxson P and Van Deventer J S J 2009 *Colloids and Surfaces a-Physicochemical and Engineering Aspects* **336** 57–63
[16] Weng L and Sagoe-Crentsil K 2007 *Journal of Materials Science* **42** 2997–3006
[17] Sagoe-Crentsil K and Weng L 2007 *Journal of Materials Science* **42** 3007–3014
[18] Rangan B V, Hardjito D, Wallah S E, Sumajouw D M J 2005 *Geopolymer: Green Chemistry and Sustainable Development Solutions*
[19] Altun N E 2014 *Fuel Processing Technology* **128** 461–470
[20] Roohbakhshan A and Kalantari B 2012 *Electron. J. Geotech. Eng.* **17** 1537–1544
[21] Prošek Z 2017 Utilization of the waste from the marble industry forapplication in transport infrastructure: Mechanical properties of cement pastes. *IOP Conf. Ser. Mater. Sci. Eng.* ***236*** 012092
[22] Chavhan P J, and Bhole, S D 2014 *Int. J. Eng. Res. Appl.* **4** 377–381
[23] Sakalkale A D, Dhawale G D, and Kedar RS 2014, *Int. J. Eng. Res. Appl.* **4** 44–50
[24] Raghvendra; Trivedi M 2017 *Int. J. Res. Appl. Sci. Eng. Technol.* **5** 712–717
[25] Tunc E T 2019 *J. Environ. Manag.* **231** 86–97
[26] Raja M A, Rajalakshmi S, Valliappan S and Dhivya K S 2017 *Int. Res. J. Eng. Technol.* **4** 1705–1708

[27] Ulubeyli G C, Turhan Bilir T and Artir R 2016 *Procedia Eng.* **161** 543–548
[28] Lee W H, Lin K L, Chang T H, Ding Y C and Cheng T W 2020 *Polymers* **12**(9) 1924
[29] Thakur A K, Pappu A and Thakur V K 2019 *Journal of Cleaner Production* **230**
[30] Osman B A A 2019 *Developing a novel geopolymer using marble dust*, A Thesis Submitted to the Faculty of Engineering at Cairo University In Partial Fulfillment of the Requirements for the Degree of Master of science In Structural Engineering, Faculty of engineering, Cairo University Giza, Egypt
[31] Tekin I 2016 *Constr. Build. Mater.* **127** 607–617
[32] Okoye F N, Durgaprasad J and Singh N B 2015 *Constr. Build. Mater.* **98** 685–691
[33] Hardjito D, Wallah S E, Sumajouw D M and Rangan B V 2005 *Aust. J. Struct. Eng.* **6** 77–84
[34] Rovnanik P 2010 *Constr Build Mater* **24** 1176–83
[35] Vijai K, Kumutha R and Vishnuram B G 2010 *International Journal of the Physical Sciences* **5**(9) 1419–1423
[36] Neupane K, Sriravindrarajah R and Baweja D 2015 *Construction and Building Materials* **94** 241–248
[37] Fernández-Jiménez A and Palomo A 2005 *Micropor. Mesopor. Mater.* **86**(1-3) 207–214
[38] Jozić D, Tiblijas D Zorica S and Bernstorff S 2012 *Insitu saxs / waxs study of the developing process of geopolymer structures*, in *ECCM15 - 15th European conference on composite materials*, Venice, Italy
[39] Tchadjié I N, Djobo J N Y, Ranjbar N, Tchakouté H K, Kenne B B D, Elimbi A and Njopwouo D 2016 *Ceram. Int.* **42** 3046–3055
[40] Krol M, Minkiewicz J, and dan Mozgawa W 2016 *Journal of Molecular Structure* 200–206
[41] Djobo J N Y, Elimbi A, Tchakouté H K and Kumar S 2016 *J. Mater. Sci.* **51** 10301–10317
[42] Hajimohammadi A, Provis J L and van Deventer J S 2011 *Journal of Colloid and Interface Science* **357** 384 – 392
[43] Yunsheng Z, Wei S, and Zongjin L 2010 *Applied Clay Science* **47**(3-4) 271–275
[44] Liew Y M, Kamarudin H, Mustafa Al Bakri A M, Bnhussain M, Luqman M, Khairul I, Ruzaidi C M and Heah C V 2012 *Construction and Building Materials* **37** 440–451
[45] Spătaru I 2016 *Contribuții privind obținerea de geopolimeri prin valorificarea unor reziduuri de producție PhD Thesis*, Facultatea Dunărea de Jos Galați
[46] Phair J W and Van Deventer J S J 2002 *International Journal of Mineral Processing* **66** 121–143
[47] Robayo-salazar R A, Mejía R, Gutiérrez D and Puertas F 2016 *Appl. Clay Sci.* 132–133, 491-497
[48] Rovnank P, Bayer P and Rovnankova P 2013 *Constr. Build. Mater.* **47** 1479–1487
[49] Puertas F and Fernandez-Jimenez A 2003 *Cem Concr Compos* **25** 287–292
[50] Tuchman D P 1992 *Department of the Interior, Bureau of Mines*
[51] Mollah M Y A, Hess T R and Cocke D L 1994 *Cement and Concrete Research* **24** 109–118
[52] Vempati R K, Rao A, Hess T R, Cocke D L and Lauer H V 1994 *Cement and Concrete Research,* **24**(6) 1153–1154
[53] Nergis, DDB, Vizureanu P, Sandu AV, Nergis, DPB and Bejinariu C 2022 *Materials* **15(1)** 202. DOI: 10.3390/ma15010202.
[54] Yahya Z, Abdullah MMAB, Ramli NM, Burduhos-Nergis DD and Abd Razak R 2018 *IOP Conf. Ser.: Mater. Sci. Eng.* **374** 012068. DOI: 10.1088/1757-899X/374/1/012068.
[55] Javadi Toghchi M, Campagne C, Cayla A, Bruniaux P, Loghin C, Cristian I, Burgnies L and Chen Y 2019 *Synthetic Metals* **251** 75-84. doi: 10.1016/j.synthmet.2019.03.026.
[56] Tadesse MG, Nagy L, Nierstrasz V, Loghin C, Chen Y and Wang L 2018 *Materials* **11(12)** 2466. doi: 10.3390/ma11122466.

IOP Conf. Series: Materials Science and Engineering 1251 (2022) 012003 doi:10.1088/1757-899X/1251/1/012003

Research on the exposure of workers in a courier company to the exhaust emissions of diesel engines

M Haiducu[1], R A Ștepa[1], E R Chiurtu[1] and I P Scarlat[1]

[1]Laboratory of Chemical and Biological Risks, The National Research and Development Institute on Occupational Safety - I.N.C.D.P.M. "Alexandru Darabont", 35A Ghencea Blv., Sector 6, 061692 Bucharest, Romania

E-mail: mariahaiducu@yahoo.com

Abstract. The paper presents the research on the development of a method for analysis of elemental carbon from diesel emissions in the workplace based on existing methods and studies on the exposure of workers in a courier company to emissions of diesel engines in the context of occupational exposure limit in Directive (EU) 2019/130 on the protection of workers from the risks related to exposure to carcinogens or mutagens at work. It is estimated that more than 3.6 million workers in Europe are exposed to diesel emissions. Occupational exposure takes place mainly in mining, construction, transport, agriculture and other activities using diesel vehicles and equipment. New data on the carcinogenicity of diesel engine emissions and the change in their classification from potentially carcinogenic (Group 2B) to carcinogenic (Group 1), as well as the high number of workers exposed to these types of chemical pollutants, highlighted the need to set occupational exposure limits for emissions diesel engines in the work environment.

1. Introduction

Most substances used at work have harmful effects on health of workers, including those exposed accidentally or because of the chemicals migrating to them from nearby sources. Concern about the carcinogenic potential of diesel emissions comes mainly from epidemiological studies on worker exposure. A study by the US National Cancer Institute and the US National Institute for Occupational Safety and Health, published in 2012, showed an increased risk of death from lung cancer among miners exposed underground to diesel engine emissions [1]. Also in 2012, the International Agency for Research on Cancer (IARC) classified diesel engine emissions as carcinogenic to humans (Group 1). It is estimated that more than 3.6 million workers in Europe are exposed to diesel emissions. Occupational exposure takes place mainly in mining, construction, transport, agriculture and other activities using diesel vehicles and equipment [2,3]. Diesel emissions are a complex mixture of gaseous substances and solid particles. The exact composition depends on the type, age and degree of wear of the engine, but also on the properties of the fuel used. The gas phase contains nitrogen, oxygen, carbon dioxide, water vapor, nitrogen oxides, carbon monoxide, sulfur dioxide and various organic compounds. The solid phase contains elemental carbon, organic compounds, sulphates, nitrates and traces of metals or other chemicals [3,4]. Measurement of exposure to diesel emissions may include monitoring carbon, nitrogen and sulfur oxides for which there are occupational exposure limit values. Respirable or inhalable particles can also be measured indicatively and compared with existing limit values. However, the measurement of these components is not sufficient to characterize the exposure to diesel emissions, in what they have the most dangerous: their carcinogenic part. The lack of occupational exposure limit values for diesel emissions and of a dedicated method for measuring them has been, until recently, a shortcoming in managing related risks in the workplace. New data on the carcinogenicity of diesel engine emissions and the change in their classification, by IARC, from possibly carcinogenic (Group 2B) to carcinogenic to humans (Group 1), as well as the high number of workers exposed [5,6]. to these types of chemical pollutants, highlighted the need to set occupational exposure limits for emissions diesel engines in the work environment. Directive (EU) 2019/130 on the protection of workers from the risks related to exposure to carcinogens or mutagens at work [7]. introduces occupational exposure limit values for 8 carcinogens or mutagens, including diesel engine exhaust emissions for which a value of 0.05 mg/m^3 was established, measured as elemental carbon. This regulation is expected to help save up to 100,000 lives over the next 50 years [8].

2. Methodology of Analysis

The method used for determining the elemental carbon from diesel engine emissions in the work environment has been developed in the PN 19 44 05 02 Project. It is based on other internationally existing methods (e.g., NIOSH 5040 method [20].) used to measure such exposures. The method measures elemental carbon (EC) from particulate matter emitted by diesel engines in the workplace atmosphere sampled on quartz filters precleaned at 800^{0}C. In addition to EC measurement, the method also allows the measurement of organic carbon (OC) and total carbon (TC). The analysis is based on the volatilization and oxidation of the carbon–containing compounds in the particles and the quantification of the carbon dioxide released, with optical correction for the pyrolysis of elemental organic carbon in the process. The method of analysis is a thermo-optical transmittance/reflectance method.

The results obtained are expressed in micrograms of carbon per square centimeter (µgC/cm^2) and allow the conversion into elemental carbon concentration expressed in milligrams per cubic meter of air (mg/m^3) depending on the flow rate and sampling time.

The CE analysis was done with ta thermo-optical OC/EC analyzer (Figure 1) after its quartz furnace has been cleaned with helium to remove ambient air. The furnace temperature was gradually increased to 870°C, thermally desorbing the organic compounds. These organic carbon (OC) gases then flew into a manganese dioxide (MnO_2) oxidizing oven, where they were quantitatively oxidized to carbon dioxide (CO_2). The CO_2 was then driven by a helium stream and mixed with hydrogen before entering a nickel catalyzed oven, to be quantitatively reduced to methane. Methane was then measured using a flame ionization detector (FID). The temperature of the sample oven is then reduced to around 500°C. At this point of the analysis, the measured carbon represents a portion of the original organic compounds on the filter sample. What remains on the filter is both original EC and EC produced from thermal pyrolysis (carbonization) of organic compounds, such as high molecular weight OCs and polymeric materials, especially natural products [9,10].

Figure 1. OC/EC analyzer with FID detector for determination Organic Carbon (OC), Elemental Carbon (EC) and Total Carbon (TC) by thermal/optical analysis (Project PN 19 44 05 02).

Sampling the air of the workplace was performed with a flexible tube personal sampling pump to which an impactor/cyclone with pretreated quartz filters. The cyclone is attached with a clip to the worker's breathing area (hemispherical area, located at the level of the worker's face, with a radius of 0.3 m). The sampling flow was set at 2 l/min. Calibration was performed for the entire sampling assembly. The sampling time was 2-2.5 h, correlated with the activity of the monitored workers varies depending on the pollution power of the source, in order to avoid overloading the filter. The sampling volume must be a minimum of 142 L. and a maximum of 19,000 L [11].

After transporting the samples to the laboratory, a representative portion of the filter (sample) was cut using a 1x1.5 cm^2 precision punch and a long needle to remove the filter portion from the punch. The analytical parameters for thermo-optical analysis are presented in Table 1. Analyzer results reported in units $\mu gC/cm^2$.

Table 1. Set of temperature points and time duration for EC analysis in the work environment.

Phase	1	2	3	4	5	6	7	8
Carrier gas	He	He	He	He	He/O_2	He/O_2	He/O_2	He/O_2
Temperature (°C)	250	500	650	850	650	750	850	940
Duration (s)	60	60	60	90	30	30	30	120

One field control filter, identical to the sampling filter, was prepared, transported to and analyzed for EC and OC in the same way as sample filters.

To determine the EC concentrations of diesel emissions the results from the analyzer, expressed in $\mu gC/cm^2$, were converted to mg/m^3, using the flow rate and sampling time, according to the equation (1):

$$C = cCE{**}0{,}001{*}A / V \quad (1)$$

Where: C = elemental carbon concentration, in milligrams per cubic meter of air (mg/m^3).

cCE = the measured value of elemental carbon on the filter sample, in $\mu g/cm^2$, multiplied by 0.001 to obtain a value expressed in mg/cm^2.

A = exposed filter area, in cm^2

V = the volume of air sampled, m^3, V = flow*time.

For the volume obtained at the pressure and temperature during sampling the correction was made at 20°C and 101.3 kPa.

The results of the thermo-optical analysis expressed in $\mu gC/cm^2$ relative to the volume of air sampled was time weighted using the corresponding exposure durations and the daily working hours. This allowed them to be compared with the mandatory limit value of occupational exposure for 8 hours provided in Annex no. 1 of the Romanian Government Decision no. 1218/2006 amended by GD no. 53/2021 [12,13], that adopted the EU limit [7].

The time-weighted average value is calculated according to the equation (2):

$$c_e = \Sigma c_i t_i / \Sigma t_i \quad (2)$$

c_e - concentration equivalent to daily exposure (mg/m^3),
c_i - the concentration measured for the working phase (mg/m^3),
t_i – exposure time corresponding to the concentration c_i (hours).

3. Material and Methodology of Research

Within the PN 19 44 05 02 project, carried out in the period 2019-2022, determinations were made in the work environment to evaluate the exposure of workers in a storage and sorting hall of a courier company in Bucharest. Five main groups of workers were identified in the hall: supervisors, sorters, forklift drivers, uploaders, unloaders. After a preliminary analysis of these groups, a number of 10 homogeneous groups of workers were defined, based on the characteristics of their activities (duration, operations, location in different areas of the hall, sources of exposure). The main groups and the corresponding homogeneous groups are indicated in table 2. One representative for each homogeneous group was designated for the personal monitoring.

Sampling was performed during the activity of unloading/sorting/loading/relocating parcels, during periods of maximum activity for curriers, as regards the time of the year and hours of the day: the second and third decade of December 2021, during early morning and late evening, respectively. A first set of samples was taken between 7.30 pm and 9.30 pm, during the unloading activity of the packages from trucks with diesel engines, and a second set was made between 8.00 am and 10.00 am during the activity of loading into vans with diesel engines. Temporary sorting and storage activities took place in parallel with the loading and unloading activities. As the activity was carried out with the doors of the hall open to allow loading or unloading in trucks, the average temperature in the hall was between 3.8 and 4^0C. The ventilation was exclusively natural, with open doors on both sides of the hall.

Sampling was performed using portable air sampling systems, on 37 mm quartz filters, with a flow rate of 2 liters/min.

The analysis was performed according to the methodology developed in the project using an OC/EC analyzer with FID detector for determination Organic Carbon (OC), Elemental Carbon (EC) and Total Carbon (TC) by thermal/optical analysis (Figure 1).

The results obtained show that the determined average values, weighted over time for 8 hours according to SR EN 689 + AC: 2019 [14], do not exceed the 8-hour occupational exposure limit value set out in Annex 4 of GD 1218/2006 [12], amended and supplemented by GD 53/2021 [13], for the ten monitored groups (Table 2).

Table 2. Results of EC determinations of diesel engine emissions for 10 homogeneous groups of workers in a storage and sorting hall of a courier company, compared to the occupational exposure limit (OEL).

Number and type of homogeneous group	Diesel engine exhaust emissions measured as elemental carbon (mg/m^3)	OEL
1-Forklift driver	0.0139	0.05
2-Sorter in the middle of the hall 1	0.0105	
3- Sorter in the middle 2	0.0063	
4-Unloaders	0.0034	
5-Supervisor in the middle	0.0015	
6-Uploaders	0.0011	
7-Sorter in the middle 3	0.0011	
8-Supervisor near offices	<0.0010	
9-Sorter in the back	<0.0010	
10-Supervisor in the back	<0.0010	

For all the 10 homogeneous groups of workers selected the results of exposure show that:

- 20% of workers were exposed to concentrations 20-30% of OEL,
- 10% exposed to concentrations 10-20% of OEL,
- 40% exposed to concentrations 2-10% of OEL,
- 30% exposed to concentrations lower than the detection limit of the method (0.001 mg/m^3).

Also, the results show that the highest exposure was for forklift drivers, which do not operate in a fix place and could be more often in areas with higher exposure. For other groups, positioning their places and the operational rules (vehicle stop their engines or use it as less as possible) reduce the exposure. The least exposure corresponds to supervisors (possibly mainly b due to a shorter exposure time) and sorters who have jobs at greater distances from the doors where the packages are loaded and unloaded in trucks and vans.

4. Conclusions

The results of the measurements show that no OEL exceedances were registered. However, there is a long-term exposure, for most of the 10 examined groups. More measurements are needed in order to establish a reliable level of exposure for each group. Reconsidering the way homogeneous groups were defined here could also be advised, now that measured data is available (for example merging some of the initial groups). The preventive and protective measures should be prioritized considering the level of measured exposure. However, because the exact composition of the carcinogens in the diesel engine emissions is unknown, measures should be taken even for the lowest exposure levels, in case some of the components are genotoxic carcinogens.

Monitoring of workers should be carried out regularly, including during various periods, when the temperature in the hall is very hot, or when the wind direction enhances emissions flow towards the hall. Shorter duration samples could help identify higher episodes of exposure, which may count, especially in case of genotoxic compounds in the emissions. Also, measures should be extended to other workers like drivers, maintenance personnel etc.

Given that the study was conducted in late 2021, during the COVID-19 pandemic, it should also be considered that diesel exhaust particles can increase both the contagion (as virus carriers) and the mortality by aggravating the pulmonary inflammation caused by SARS-CoV-2, by substances with similar mechanisms [16,17].

Diesel engine emissions are proven to be dangerous to health, mainly due to – but not limited to - their carcinogenic effects. Workers in enclosed spaces, likes mines, are likely to be considerably exposed to such emission. However, other type of workers may be subject to hazardous diesel exhaust exposure, too. It is, therefore important to properly monitor exposures and to centralize the data obtained in order to check and track the effect over time of these exposures on the health of exposed workers [18].

Identifying, analyzing and assessing the risks of injury and occupational disease are the basic tools of risk management to reduce and eliminate by technical and organizational measures. The benefit of risk assessment allows identification of the problems faced by workers, which requires the adoption of preventive measures [19-22].

The prolonged exposure to diesel engine exhaust (DEE) can induce a neuroinflammatory response in the brain as suggested by rat studies. The brain is a relevant target for the effects of inhaled DEE and the long-term exposure may promote the development of Alzheimer's disease [23,24].

Acknowledgement

This work was funded by the MCID, Project PN 19 44 05 02, "Development of a method for analysis of elemental carbon from Diesel engine exhaust emissions in the workplace in the context of the amendment of Directive 2004/37/EC on the protection of workers from the risks related to exposure to carcinogens or mutagens at work", 2019-2022.

References

[1] Coble J B, Stewart P A, Vermeulen R, Yereb D, Stanevich R, Blair A, Silverman D T and Attfield M 2010 *Ann. Occup. Hyg.* **54** 747–761.

[2] Taxell P and Santonen T 2017 *Toxicological Sciences* **158(2)** 243–251.

[3] McClellan R O, Hesterberg T W and Wall J C 2012 *Regulatory Toxicology and Pharmacology* **63** 225–258.

[4] McDonald J D, Barr E B, White R K, Chow J C, Schauer J J, Zielinska B and Grosjean E 2004 *Environ Sci Technol* **38** 2513–2522.

[5] Benbrahim-Tallaa L, Baan R A, Grosse J et al. 2012 *Lancet Oncol* **13** 663–664.
[6] Attfield M D, Schleiff P L, Lubin J H, Blair A, Stewart P A, Vermeulen R, Coble J B and Silverman D T 2012 *J Natl Cancer Inst* **104.**
[7] Directive (EU) 2019/130 of the European Parliament and of the Council on the protection of workers from the risks related to exposure to carcinogens or mutagens at work.
[8] European Commission, Proposal for a Directive of the European Parliament and of the Council amending Directive 2004/37/EC on the protection of workers from the risks related to exposure to carcinogens or mutagens at work/Impact Assessment, 2016.
[9] Organic Carbon/Elemental Carbon (OCEC) *Laboratory Instrument Manual, Sunset Laboratory Inc.*, 2018.
[10] SR CEN/TR 16243:2016 Ambient air quality - Guide for the measurement of elemental carbon (EC) and organic carbon (OC) deposited on filters.
[11] Method 5040, Diesel Particulate Matter (as Elemental Carbon), NIOSH**,** 2003.
[12] Government Decision no 1218/2006 laying down minimum requirements for the health and safety at work for the protection of workers from risks related of chemical agents.
[13] Government Decision no 53/2021 for the modification of the Government Decision no. 1218/2006, as well as for the modification and completion of the Government Decision no. 1093/2006.
[14] SR EN 689 + AC: 2019 Workplace exposure - Measurement of exposure by inhalation to chemical agents - Strategy for testing compliance with occupational exposure limit values.
[15] SR EN 14530:2004 Workplace atmospheres - Determination of diesel particulate matter - General requirements.
[16] Comunian S, Dongo D, Milani C and Palestini P 2020 *Int. J. Environ. R. Public Health* **17**.
[17] Poyraz B M, Engin E D, Engin A B and Engin A 2021 *Environmental Toxicology and Pharmacology* **86**.
[18] Haiducu M, Ştepa R A, Chiurtu E R and Schmitz-Felten E 2019 *EEMJ* **18** 825–830.
[19] Buică G, Antonov A E, Beiu C, Pasculescu D and Risteiu M 2018 *SIMPRO.*
[20] Sioustis IA, Axinte M, Prelipceanu M, Martu A, Kappenberg-Nitescu DC, Teslaru S, Luchian I, Solomon SM, Cimpoesu N and Martu S 2021 *Applied Sciences* **11(9)** 3824.
[21] Cimpoesu N, Stanciu S, Vizureanu P, Cimpoesu R, Achitei DC and Ionita I 2014 Journal of mining and metallurgy section B-metallurgy **50(1)** 69-76. DOI: 10.2298/JMMB121206010C.
[22] Yahya Z, Abdullah MMAB, Ramli NM, Burduhos-Nergis DD and Abd Razak R 2018 *IOP Conf. Ser.: Mater. Sci. Eng.* **374** 012068. DOI: 10.1088/1757-899X/374/1/012068.
[23] Gerlofs-Nijland M E, Van Berlo D, Cassee F R, Schins R P F, Wang K and Campbel A 2010 *Particle and Fibre Toxicology* **7(12**) 2010.
[24] Hullman M, Albrecht C, Van Berlo D, Gerlofs-Nijland M E, Wahle T, Boots A W, Krutmann J, Cassee F R, Bayer T A and Schins R P F 2017 *Particle and Fibre Toxicology* **14**(35).

IOP Conf. Series: Materials Science and Engineering 1251 (2022) 012004 doi:10.1088/1757-899X/1251/1/012004

High-resolution NMR and MALDI-MS molecular profiling of craft beers

C Siciliano and D Procopio

Dipartimento di Farmacia e Scienze della Salute e della Nutrizione,
University of Calabria, I-87036 Arcavacata di Rende, Italy

E-mail: carlo.siciliano@unical.it

Abstract. Beer is a complex aqueous mixture of organic and inorganic compounds, such as carbohydrates obtained from barley or wheat malts, hops and yeast, ethanol, proteins, vitamins, and minerals. Phenols and their derivatives obtained by chemical interactions with biogenic amines, together superior alcohols and amino acids, are the other compounds present in minor proportions in beer, Phenolic components possess high antioxidant activities, and regulate the oxidative stability of beer, although bitterness of this beverage originates from hops. Since these antioxidants can produce beneficial effects on the human health after consumption, it is challenging to have in hand the molecular profiles of beers, in order to investigate their contents in bioactive compounds. Today, a significant number of small producers have strongly invested in craft beers, produced by using different hop blends, starting materials that are rich in phenolic antioxidants. The present work discloses an instrumental analytical method which enables the rapid obtainment of the molecular profiling of craft beers, with the assistance of selective extractions made by using multi-walled carbon nanotubes (MWCNTs). In this investigation, high-resolution ^{1}H NMR, and LDMS and MS/MS analysis were applied for the "beeromic" analysis of a craft beer brewed in Calabria (the Southern part of Italy). The spectral data, confirmed that this kind of beer predominantly contains the hop bitter metabolites humolones and lupulones, also highlighting the presence of phenols and phenolamides which are known for their bioactivity against cancer. A profile of the principal volatile hop components was also obtained by GC-MS.

1. Introduction

1.1. Molecular profile of a craft beer

Beer is an alcoholic beverage, brewed with barley, hops, water, and yeast. The chemical composition of beer varies considerably and consists in a very complex molecular mixture, where their components exert a crucial role in defining the quality and peculiar characteristics of beer [1]. The most interesting organic compounds in beer are polyphenols, and other antioxidants having different molecular structures [2].

Craft beers are produced in local microbreweries, represent a differential beverage with the respect to commercial beers and are worldwide distributed in the pubs and markets. Craft beers are known for their flavour, brew, and regional production. This type of beverage takes as raw material only malt and hops, and the brewing process does not include filtration steps and the use of chemical additives, based on specific recipes according to the preference of consumers. The molecular profile of craft beer is further defined by metabolites arising from the cereal fermentation and yeast reactions [3], and it also consists of carbohydrates, minerals, alcohol, and water [4]. In addition, craft beer is generally unfiltered, unpasteurized, and the addition of nitrogen or carbon dioxide is avoided during the brewing process [5]. Several hundred metabolites characterize the molecular profile of craft beer, which contain not only aliphatic alcohols different from ethanol, but also prenylated polyketides, aldehydes, esters, histamines, tannins, and other organic compounds [6].

Among the bioavailable organic compounds that define the chemical profile of craft beer, phenolic compounds are the most abundant [7,8]. Polyphenols are important compounds for beer quality as they can contribute to bitterness, colour, body, and astringency and can therefore influence their acceptance [9]. A high number of different polyphenols have been detected in beers, both from barley and hop [10]. Three polyphenol groups (flavan-3-ol, flavanols, and phenolic acids) are generally present in beers, and the most concentrated phenolic acid recognized are ferulic, gallic, caffeic, and p-coumaric acids [11]. However, little information is available about the molecular fingerprinting of bioactive phenolic compounds and other antioxidants in craft beer [12]. These natural compounds display bioactivities against the oxidative stress, which is the etiological cause of many diseases in humans, such as atherosclerosis and diabetes, and can provide benefit against mutagenic and carcinogenic agents [13]. Phenolic acids and polyphenols in beer have been identified as modulators of the cholesterol level [14]. Finally, these compounds give to beer stability against microbiological damage [15], and consistency of foam [16]. Because of the significant growth of the craft beer consumption, the determination of the total content and/or the individual phenolic compounds in craft beers is challenging, due to the requests of consumers for a high-quality and healthy product. Today, many efforts have been focused on the assessment of physical and chemical parameters of different types of craft beers.

1.2. Hop composition
The polyphenol content in craft beer depends on the amounts and blends of hops added during its production [17]. Hops (Humulus lupulus L.) contain several metabolites that affect quality and sensory characteristics of craft beer, also determining its chemical profile and fingerprinting. The most important bitter compounds are the α-acids (humolone, and the major homologues n-, co-, and ad-α acids), and β-acids (lupulone, colupulone, and adlupulone) (figure 1) [18]. When hops are boiled for a prolonged time, or during the storage of craft beer, the α-acids isomerize into iso-α-acids with cis and trans configurations, which sensibly modify the bitterness [19-21]. Their congeners, β-acids, are essential oils, responsible for the typical hoppy aroma of craft beer, and tend to evaporate over time, modifying the organoleptic characteristics of craft beer [22].

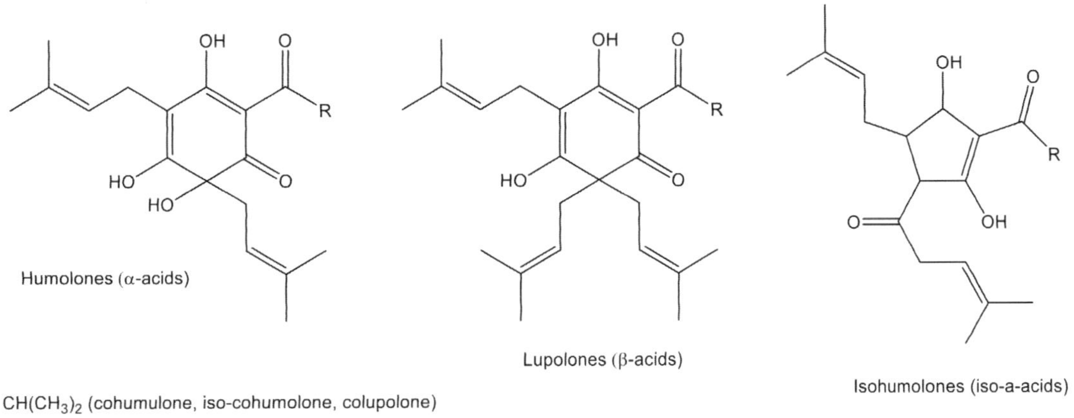

R: $CH(CH_3)_2$ (cohumulone, iso-cohumolone, colupolone)
$CH_2CH(CH_3)_2$ (humolone, iso-humulone, lupulone
$CH(CH_3)CH_2CH_3$ (adhumulone, iso-adhumulone, adlupulone)

Figure 1. Structures of the most common bitter acids in hops.

1.3. Instrumental analysis for determining the chemical profile of beers

Because of the interest in health benefits of natural flavonoid and prenylated antioxidants, there is a growing demand for accurate, reproducible and sensitive analytical methods to quantify these compounds in beer. Several instrumental analytical techniques have been used to investigate the molecular profile of beer in "beeromics" approaches [23], to identify natural metabolites derived from the hop plant and flowers, as specific and potential targets of biochemical investigation of beer. In the past decade, fast and automated isolation-separation-identification techniques have been developed for the chemical characterization of the antioxidant profile of commercial beers [24]. The early determinations of phenolic compounds and bitter acids in commercial beer have been made by colorimetric methods [25], thin layer chromatography (TLC) [26], capillary electrophoresis [27]. However, all methods showed very low resolution, and the unequivocally identification of single compounds remained uncertain. The profile of metabolites in natural complex matrices can successfully be established by ultrafast high-performance liquid chromatography coupled to tandem mass spectrometry (UHPLC-MS/MS) methods, as well as by linear and tandem mass spectrometry (MS and MS/MS) experiments, operating under electrospray or matrix assisted laser desorption ionization (ESI and MALDI mass spectrometry, respectively) [28-30]. To analyze beer samples, the coupling of HPLC with mass spectrometry (MS) using electrospray ionization (ESI) [31] or atmospheric pressure chemical ionization (APCI) [32] has previously been reported. In the case of hop extracts, chromatographic separation coupled to mass spectrometric detection array, in parallel with gas-chromatography/mass spectrometry methods (GC-MS), have advantageously been applied for the analysis of the essential oil and the volatile fraction of extracts [33]. Direct thermal desorption (DTD) [34] and headspace solid phase microextraction (HS-SPME) [35] have also been employed. The fingerprinting of hop oil and the profile of bitter acids have been investigated by hyphenated chromatographic-mass spectrometric techniques (RP-HPLC-MS, UHPLC-MS/MS) [36], while the organoleptic characteristics of beer have been defined by gas chromatography-olfactometry (GC-O) [37]. Analytical methods based on RP-HPLC coupled to ultra-violet (UV) [38], electrochemical [39] or evaporate light-scattering [40] detection apparatus have been introduced. Furthermore, gas chromatography and electromigration techniques, such as capillary zone electrophoresis or capillary electrochromatography [41], have been employed as well. Nevertheless, none of these methods allow the rapid and simultaneous determination of a wide range of compounds and therefore cannot definitively provide the molecular profile of an extremely complex matrix as beer. High-throughput analytical instrumentation based on high mass accuracy measurements and tandem mass-spectrometry (MS/MS experiments) can successfully be applied to generate fragmentation patterns useful for the structural elucidation of isobaric bitter acids, the detection of β-bitter acids and their transformation products in commercial and craft beer [42].

1.4. Nuclear Magnetic Resonance (NMR) techniques

The structures of α- and β-acids from hop have recently been elucidated by complementary instrumental analysis based on 1D/2D high-resolution NMR in solution [43]. This study demonstrates the advantages of the use of high-resolution instruments together with accurate mass measurements, often supported by chemometrics. NMR spectroscopy represents the most advantageous instrumental analytical and direct tool in obtaining the molecular profile of natural complex matrices of vegetable and animal origin [44-46]. Today, metabolite fingerprinting by NMR methods is a fast-growing field which is increasingly applied for quality control of foods and herbal products [47]. With reference to beer, bitterness, autoxidation reactions, changes during aging, and other peculiar properties can been investigated by high-resolution NMR techniques [48-52]. The main components of the α- and β-acid families of hop have been identified, and their structures were confirmed by high-resolution mass spectrometry and MS/MS experiments. Beer samples can straightforwardly be analyzed by NMR without any pre-treatment, avoiding purification and preconcentration steps, or chemical manipulation for the pre-derivatization of metabolites, and preserving loss of low concentrated metabolites.

NMR spectroscopy might provide exhaustive information on the chemical fingerprinting of commercial and craft beer. This instrumental tool is simple, fast, reliable, fully reproducible, and preserves sample components from deterioration. The qualitative and quantitative data provided by NMR are better than those provided by all other instrumental methods cited before, because this spectroscopy represents a primary analytical method, with the intensity of the signals directly correlated to the molar concentration of the identifiable metabolites, with no need for single calibration curves calculated for all compounds under evaluation. High-resolution NMR spectra might provide not only the unequivocal assignment of major hop metabolites, but spectral data also highlight isomerization and degradation products of hop bitter acids generated during brewing or storage and aging of beer [53]. Despite the evident advantages that NMR techniques can offer, spectroscopic information available on bitter compounds present in craft beers are limited, and today only such a few articles have previously been published dealing with the qualitative/quantitative investigation of the bitter acid profile of craft beer [54]. Nevertheless, signal overlapping represents a common drawback that might hinders metabolites identification, resulting in an incomplete data interpretation.

1.5. Beer analysis by extraction

Most methods for beer analysis have used degassing to prevent bubbling during pipetting, followed by sample filtration prior to injection. Despite a relatively simple sample preparation, very different and low recoveries of the target antioxidant metabolites and bitter acids from beer could result [55]. Solid phase extraction (SPE) [56], cloud point extraction (CPE) [57], and magnetic mixed hemi-micelles SPE (MMHSPE) [58] have been employed. Chemical interferences with carbohydrates, and physical interactions with the solid phases used for mechanical filtrations of samples can limit the effectiveness of processes in terms of extraction yields, determining the underestimation of concentrations and severe losses of low abundant metabolites.

The use of carbon materials, such as graphene and nanotubes, can facilitate the extraction of metabolites from complex matrices [59-62], allowing the selective recovery of whole classes of organic compounds based on the electronic interactions between the carbon structure of the extractive solid medium and the molecular structures of metabolites. Extractions with these carbon materials could better work by simply choosing the ideal organic solvent systems [63,64]. In this context, multi-walled carbon nanotubes (MWCNTs) have successfully been proposed [65].

2. Experimental details

2.1. Chemicals and materials

Craft beer was commercially available in local markets. All chemicals, solvents, and NMR deuterated media (D_2O, $CDCl_3$ and DMSO-d_6, all solvents 99.9% isotopically pure) were purchased were purchased from Aldrich (Italy). Multi-walled carbon nanotubes (MWCNTs) were synthesized in our laboratories and used after purification, as reported elsewhere [72,73].

2.2. NMR analysis of the beer dry residue

For direct NMR spectroscopic analysis, a volume of 20 mL of beer was freed of carbon dioxide by sonication in an Erlenmeyer shaking flask, for 15 minutes. The liquid was dried at low temperature under vacuum by using a Speed-Vac apparatus, and the final slurry brown residue (5.7 mg) was completely dissolved in DMSO-d_6 (0.75 mL) and subjected to the NMR analysis.

2.3. Solvent extractions

The dry residue of craft beer (25 mg) was dissolved in D_2O (3 mL) and extracted in triplicate with cold hexane (6 mL total volume). The organic phase was dried over Na_2SO_4, filtered on a paper disk, and evaporated to dryness, without using vacuum systems. The oily residue (5.3 mg) was dissolved in $CDCl_3$ and subjected to the NMR and SPE-GC-MS analysis. The aqueous phase was treated with MWCNTs (2 mg) in the presence of ethyl acetate (2 mL), under magnetic stirring for 2 hours at room temperature. The carbon material was recovered over a paper filter and washed three times with fresh ethyl acetate (6 mL total volume). The organic solution was dried over Na_2SO_4, filtered over a paper disk, and the solvent was removed by a rotary evaporator. The solid residue (8.7 mg) was dissolved in $CDCl_3$ and analyzed by NMR. The aqueous layer was re-treated with $CHCl_3$ (2 mL) in the presence of MWCNTs (2 mg), under magnetic stirring at room temperature, for 2 hours. The carbon material was filtered off and washed three times with fresh chloroform (6 mL total volume). The organic phase was dried over Na_2SO_4, filtered on a paper disk and evaporated to dryness under vacuum. The oily residue (3.8 mg) was dissolved in $CDCl_3$ and a proton NMR spectrum was recorded. The residual aqueous phase was directly analyzed by NMR spectroscopy.

2.4. High-resolution ^1H NMR Spectroscopy
All spectra were recorded at 25 °C on a Bruker Avance 300 Ultrashield spectrometer, equipped with a 5 mm BB0 probe and a temperature control unit, and at the frequency of 300.112 MHz for the proton resonance. Chemical shift values were reported in ppm and referred to the resonance of the residual unlabelled solvents (7.21 ppm and 2.50 ppm for spectra recorded in $CDCl_3$ and DMSO-d_6, respectively). Acquisition and elaboration parameters were as previously published [74].

2.5. Mass spectrometry
Laser Desorption Mass Spectrometry (LDMS) and MS/MS experiments were performed by applying experimental and instrumental conditions as previously reported [75].

2.6. Gas chromatography-mass spectrometry
Experimental conditions were as follow: an Agilent Model 6850 gas chromatograph was used. The instrument was equipped with a flame ionization detector (FID), a capillary inlet system, a DB-23 (60m, 0.25 mm, 0.251 m) column, and a Model Maestro MPS2XL multipurpose sampler with a 10μL syringe was employed. An Envi-18 cartridge was used. The programmed oven temperature was:130°C for 1 min, 130–170°C at 6.5°C/min, 170–215°C at 2.75°C/min, 215°C for 12 min, 215-230°C at 40°C/min, 230°C for 3 min. The carrier gas (H_2) flow rate was 1.7 mL/min; constant pressure mode was setup. The split ratio was set at 50:1. The temperatures of injector and detector were 270 and 280°C, respectively.

3. Discussion
We disclose preliminary data obtained from the investigation of the molecular profiling of a commercially distributed craft beer brewed in Calabria (the Southern part of Italy). The aim of this study was to investigate the molecular profile of an American Pale Ale (IPA) beer, obtained using a blend of different kins of hops. We pointed our attention to the possible recognition of the minor components of craft beer, namely phenol derivatives and bitter acids, which have been proved to possess antioxidant bioactivity. Consequently, we have designed an approach based on the application of organic solvent extractions, assisted by multi-walled carbon nanotubes (MWCNTs). The rationale was that the employment of a specific carbon adsorbent, together with the appropriate solvent system, could enable the selective recovery of fractions enriched in antioxidants and, as desirable, the delivery of fractions containing the main compounds which determine the aroma beer. A combination of instrumental techniques was employed, the high-resolution proton nuclear magnetic resonance for the obtainment of general information about the chemical profile of each fraction obtained from MWCNTs, mass spectrometry for collecting insights on polar compounds, and gas chromatography-mass spectrometry for the profile of the beer aroma components.

We started our investigation by directly subjecting a sample of the raw beer to ^1H NMR analysis. The beer case of study was brewed in August 2021, and it was analyzed after a storage period of 30 days in the dark and at 15°C. The physical and chemical manipulation of the matrix sample was limited only to the preliminary degassing of 20 mL of beer, which was sonicated for 15 minutes in a bath maintained to the temperature of 20°C. The spectrum was recorded without water suppression, and it is displayed in figure 2.

NMR spectroscopy highlighted that the total content of water and ethanol (the presence of these components was confirmed by the strongly intense signals at 4.78 for water, and 3.54 and 1.08 ppm for ethanol), disturbed the intensities of all other minor resonances assignable to the low concentrated metabolites in beer. Upon the adopted experimental conditions, the identification of other signals in the sub-spectrum resulted to be very difficult.

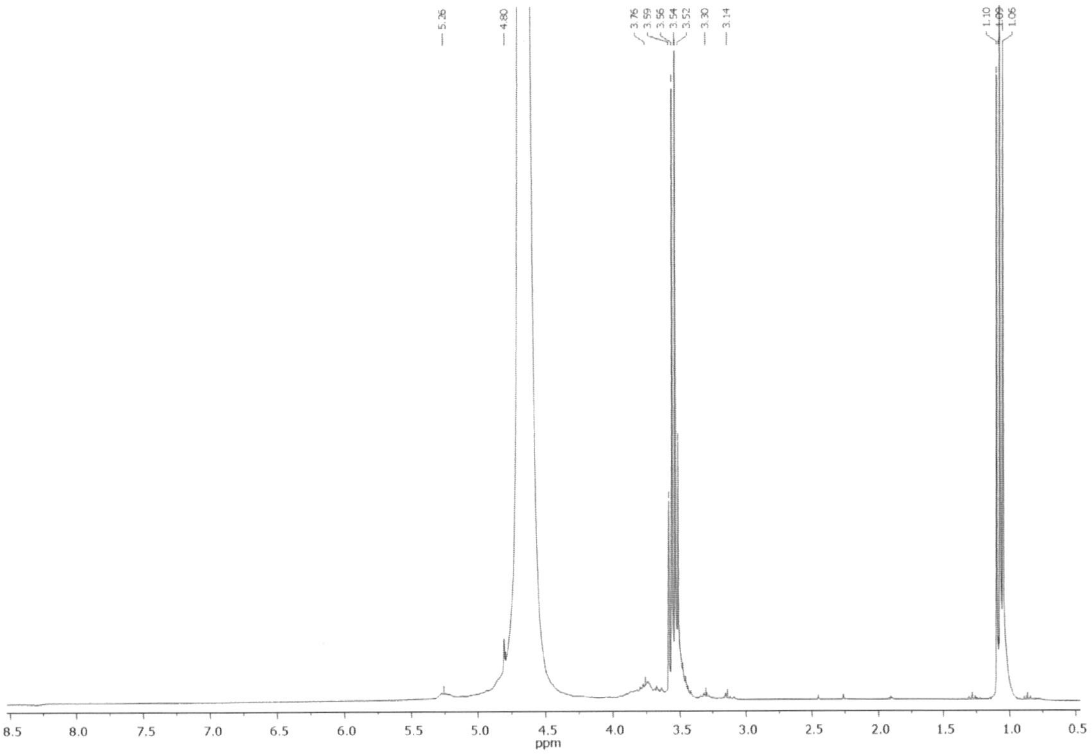

Figure 2. High-Resolution ^{1}H NMR spectrum of a sample of degassed raw craft beer, recorded without the inverse recovery water suppression pulse sequence.

The spectrum obtained by applying a pulse sequence to generate the water signal suppression, did not was helpful in identifying the diverse chemical components in the beer sample. The complexity of the plot was evident, and the ^{1}H NMR spectrum of beer shows only broad and strongly overlapping signals due to the protons of carbohydrates. Moreover, the resonances close to the suppressed frequency of water resulted to be overwhelmed or impossible to phase and correctly resolve, giving rise to distortions of the spectral baseline together with a reasonable signal reduction. Our preliminary experiments confirmed the literature evidence, highlighting that the complexity of the ^{1}H NMR of the raw beer is an important drawback hampering the analysis [66]. In order to minimize this trouble, we adopted the freeze-drying for sample preparation. In that case, an aliquot of 20 mL of the degassed beer was dried up by in an ultra-centrifuge at 4 °C, to afford a thick brown residue which was weighed and redissolved in DMSO-d_6, without pH adjustment.

The sample was analyzed by ^{1}H NMR spectroscopy, using a normal proton pulse sequence and avoiding the application of pre-saturation, relaxation cycling, or selective excitation pulse sequences. The recorded spectrum is displayed in figure 3. Chemical shifts of the resolved signals were used to identify chemical components, based on the database already reported in the literature [67,68]. No appreciable differences were found between the experimental and the literature values, Nevertheless, the determination of the chemical profile of flavour and taste of beer appeared to be critical.

We attempted an initial analysis of the spectral data, due to the richness of resonance signals featured by the spectrum. In DMSO-d_6 the most part of signals was easily attributed, subdividing the total spectrum in several well-distinguished spectral regions. As expected, all spectral windows displayed signal networks which were due to the different classes of chemical compounds in beer.

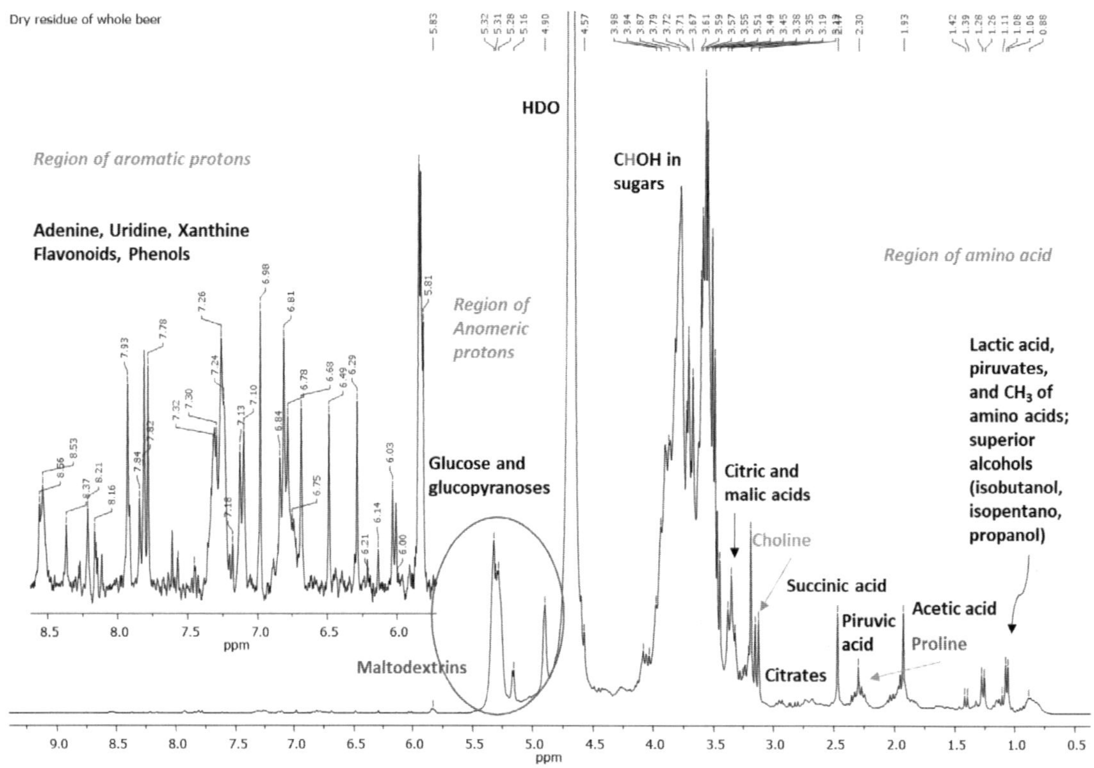

Figure 3. High-Resolution ^{1}H NMR spectrum recorded for a sample prepared from an aliquot of the craft beer dry residue.

In particular, the high-resolution ^{1}H-NMR spectrum was characterized by the presence of a low-field region (in the chemical shift interval between 5.60 and 9.0 ppm), presenting resonances due to the aromatic and olefinic protons, the most intense of which were symptomatic for the structure of xanthohumol (signals between 6.10 and 7.80 ppm), α-humulene and linalool (5.50-6.00). In the same spectral region, it can also be detected the signals of adenine, xanthine, uridine, and the family of phenol residues in flavanols. The spectral mid-low field window, in the chemical shift interval between 5.45 and 4.25 ppm, displayed signals attributable to the anomeric protons of carbohydrates (glucose and glucopyranoses) and the CH=CH protons of unsaturated free fatty acids (MUFA and PUFA). Finally, methyl, methylene, and methine resonances generated by the protons of α- and β-bitter acids, appeared up-field, between 0.80 and 3.20 ppm, however showing not-resolved signals and overlapping to the resonances of choline derivatives and carboxylic acids (citric and malic acids), which were predominant. The other carboxylic acids which can be detected in beer, piruvic and succinic acids, showed their signals as singlets at 2.30 and 2.49 ppm, respectively, while the singlet of the methyl protons of acetic acid was detected at 1.93 ppm. Lactic acid, lactates, citrates and piruvates gave signals in the intervals between 2.49 and 3.00 ppm. Ramified amino acids were detected by their side-chains methyl proton signals between 1.00 and 1.50 ppm, with the proline ring signals appearing as multiplets at 1.75-2.10 and 2.25-2.36 ppm. Propyl alcohol, isobuthyl alcohol and isopenthyl alcohol were detected by the signals appearing in the spectral sub-window between 1.00 and 1.30. In the recorded spectrum, signals attributable to fatty acids appeared at 0.90 ppm (protons of all terminal methyl groups) and 1.28 ppm (inner-chain methylene protons) were not assigned to any single fatty acid chain. Although the results highlighted the possibility of obtaining a total qualitative chemical profile of the craft beer, no definitive information about the presence of the different bitter acids was collected, indicating the need for alternative methods that could enable the selective recovery of fractions enriched in well-defined classes of organic compounds.

We further tried to obtain more information about the chemical profile of the craft beer under analysis. It has been proved [69] that the use of 2-butanol can furnish beer extracts characterized by ^{1}H NMR spectra that can allow the recognition of bitter acids. However, the signals still showed considerable overlapping, and 2-butanol is difficult to eliminate from samples. In order to fulfil our aim, we designed a method of simple extractions based on the use of solvents having different polarities, with the recovery of the classes of hop metabolites assisted by MWCNTs as the adsorbent-releasing material. The use of selective extraction by multi-walled carbon nanotubes (MWCNTs) of organic components from craft beer, followed by high resolution ^{1}H NMR analysis for metabolite profiling, has not been reported before. NMR signals were assigned using a combination of data obtained in comparison with spectra of available reference standards, and with data base already reported in the literature.

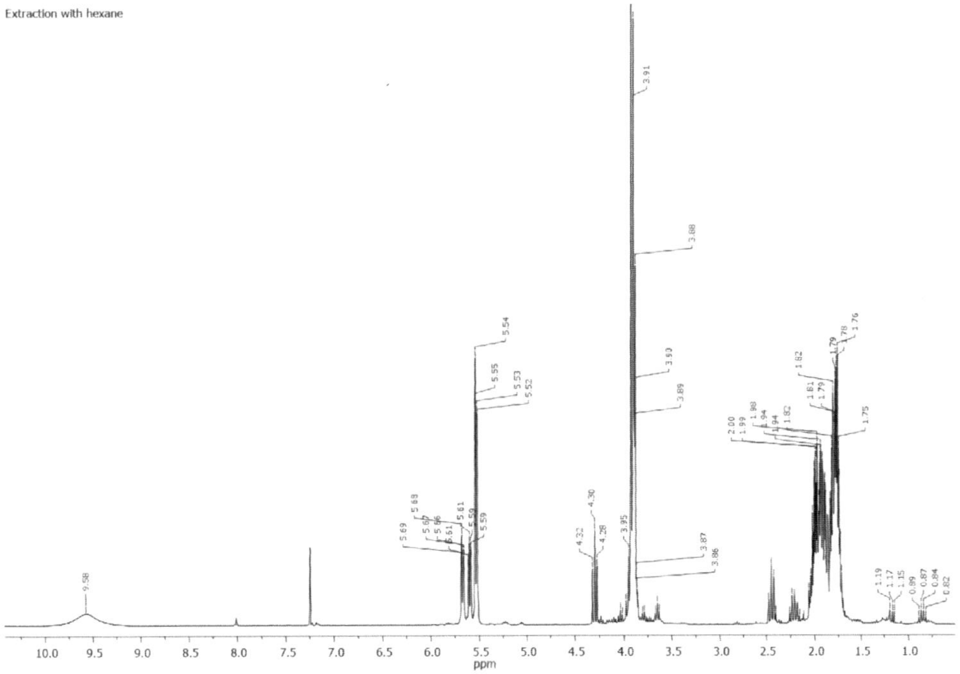

Figure 4. High-Resolution ^{1}H NMR spectrum recorded on the sample obtained by extracting the craft beer dry residue with cold hexane.

In the first step of the procedure, the dried craft beer residue was dissolved in distilled water and cold hexane was used for the solvent extraction. The extracts were dried at low temperature, re-dissolved in deuterated chloroform ($CDCl_3$), and analyzed by NMR spectroscopy, giving the spectrum depicted in figure 4. It is evident the simple pattern of signals that can be observed, and this result was totally unexpected. The spectrum showed three groups of well-resolved resonances, at 1.50-2.50, 3.5-4.20, 5.50-5.75 ppm. The broad signal centred at 9.58 ppm, together with the series of intense singlets centred around 3.86-3.95 ppm were significative of phenolic OH functions and CH_3 protons of methyl esters, while signals in the interval between 5.50 and 5.75 were attributed to vinyl proton resonances, and those at 1.50-2.50 to CH, CH_2 and CH_3 groups in aliphatic chains.

Literature database confirmed that the hexane fraction predominantly contained compounds having long chain, monoterpene, and sesquiterpene structures. We thus hypothesized that hexane was successfully employed to extract the volatile fraction of hop flower components. Despite the simplicity of the signal spectrum, suggesting the presence of only such a few organic species, the GC-MS analysis of the sample used for NMR spectroscopy, highlighted a surprising chemical profile. The obtained GC run is illustrated in figure 5, and all compounds identified are reported. Methyl and ethyl esters of long aliphatic chain carboxylic acids, monoterpenes, sesquiterpenes and some phenolic derivatives were detected.

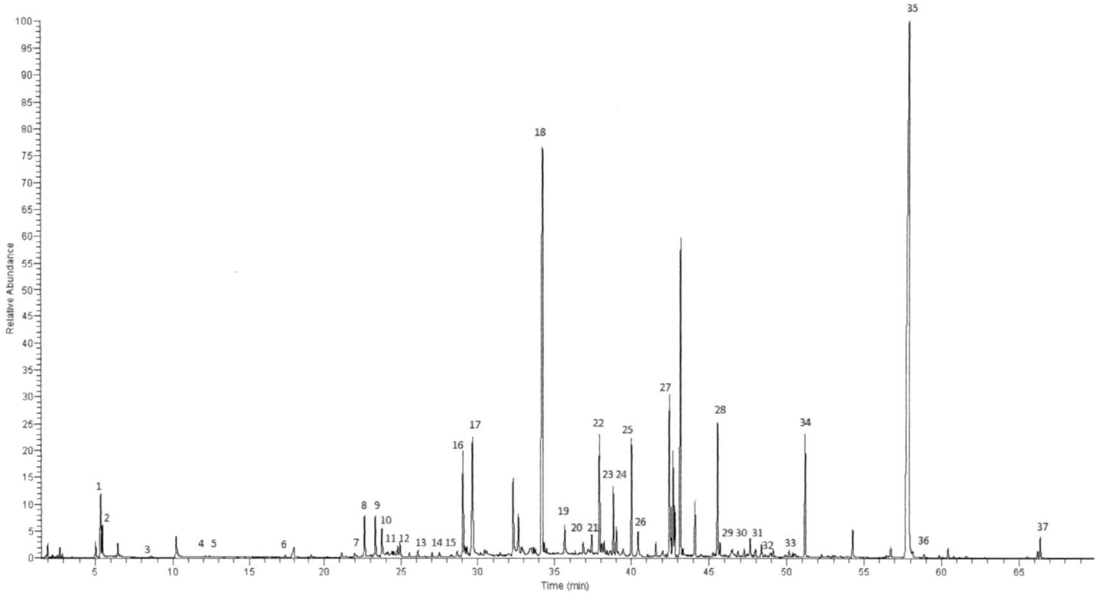

Figure 5. GC-MS analysis of the hexane extract. (1: 2-methyl-4-butanol; 2: 2-methyl-1-butanol; 3: ethyl butyrate, 4: ethyl-2-methyl-butyrate; 5: ethyl-3-methyl-butyrate, 6: isobutylether; 7: β-pinene; 8: ethyl hexanoate; 9: terpinolene; 10: p-mentha-1,3,8-triene; 11: sylvestrene; 12: methyl-4-ethyl-4-pentenoate; 13: *trans*-ocimene; 14: *cis*-ocimene; 15: ethyl 5-methyl-hexanoate; 16: linalyl isobutyrate; 17: phenylethyl alcohol; 18: ethyl octanoate; 19: 3,7-dimethyl-6-octen-1-ol; 20: *cis*-geraniol; 21: ethyl 4-methyl-octanoate; 22: 2-methoxy-4-ethyl phenol; 23: ethyl nonanoate; 24: undecanol; 25: methyl-3,7-dimethyl-2,6-octadienoate; 26: ethyl decanoate; 27: ethyl 4-decenoate; 28: cariophyllene; 29: farnesene; 30: eudesmene; 31: selinene; 32: cadinene; 33: α-calacorene; 34: ethyl dodecanoate; 35: internal standard; 36: ethyl tetradecanoate; 37: ethyl hexadecanoate).

We continued the investigation, by treating with MWCNTs the aqueous phase separated from the hexane layers, under slow magnetic stirring at room temperature, for 2 hours, after adding ethyl acetate. The carbon material was filtered off and washed with small amounts of $CDCl_3$ to prepare the sample for spectroscopy. The HR-^{1}H NMR spectrum is illustrated in figure 6.

The results highlighted that the fraction of metabolites obtained by the system MWCNTs/ethyl acetate prevalently contained the main flavonoid antioxidants, caffeic acid derivatives and bitter acids that have been recognized as the components of hop extracts. The proton spectrum (figure 6) appeared complicated by a high number of signals. However, confrontation with the literature database, allowed the easy attribution of the different resonance patterns to each family of compounds. The most intense signals were due to the quinoid structures of humulone (α-acids) and lupulone (β-acids) series, featuring isoprenyl side chains, identifiable by observing the signals in the interval between 1.35 and 1.70 ppm (for the allylic CH_3 groups), and between 2.50 and 3.00 ppm (for the allylic CH_2 protons). Based on these chemical shift intervals, the total spectrum can be divided into two broad regions, with the signals between 0.50 and 4.50 ppm significative of the structures of humulone and its analogs, adhumolone and cohumolone, and of lupulone and its analogs, adlupulone and colupulone. In this case, although all aliphatic proton resonances aided the identification of the most concentrated bitter acids, several signals displayed overlapping and/or low intensities, thus other minor components of the beer resin (e.g., the isomerized bitter acids due to long brewing treatments and/or aging) were not separable and clearly identifiable, at least upon the sensitivity limits of the NMR techniques. The second spectral region between 4.90 and 8.00 ppm showed a lower number of signals, which were due to xanthohumol, and prenylated chalcones and flavanones. Biogenic amine conjugates of caffeic acid and its methylated derivatives were also recognized. It is attractive to point the attention on signals at 5.60-5.75 ppm, and 6.60-7.50 ppm, respectively generated by the resonances of vinylic and aromatic protons of caffeolyl tyramine, a biogenic amine conjugate possessing a high anticancer bioactivity [70].

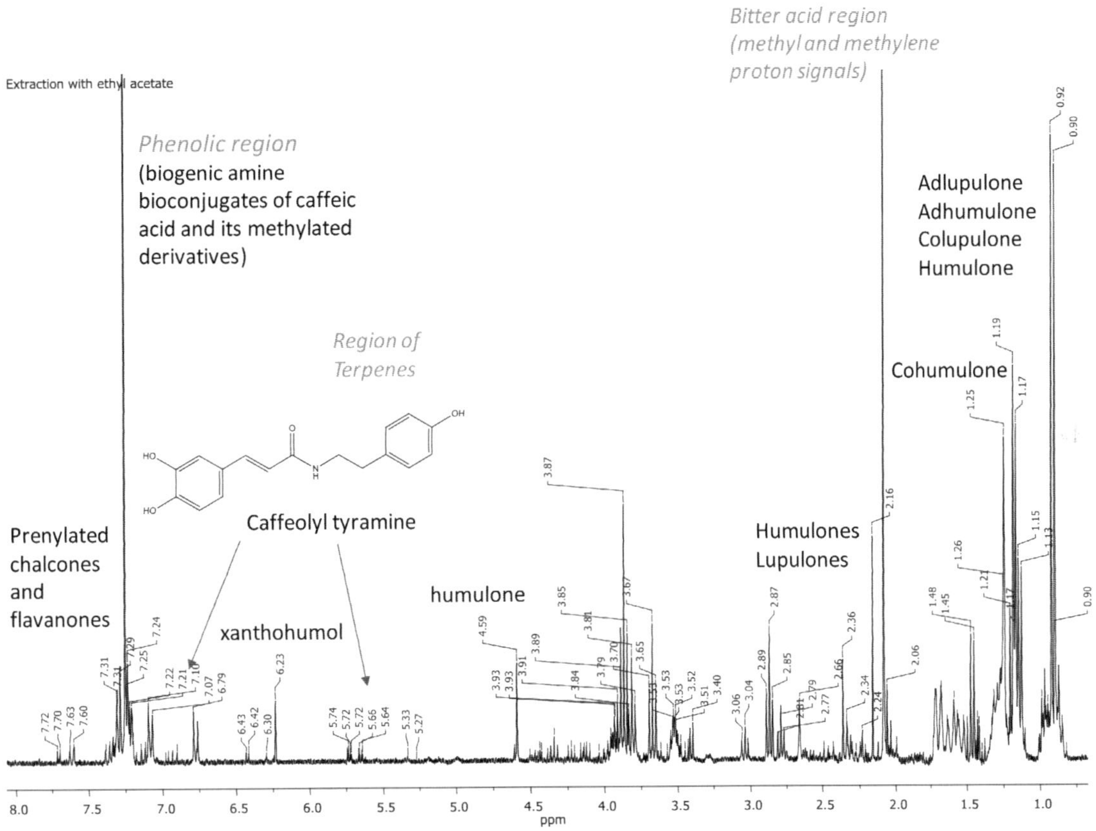

Figure 6. High-resolution ^{1}H NMR spectrum recorded on the sample obtained by the treatment with the heterogeneous system MWCNTs/ethyl acetate.

From the signals observed, and related to the sensitivity of the spectroscopic method, it was not possible to confirm the presence of isoxanthohumol, desmethylxanthohumol, 6-prenylnaringenin, 8-prenylnaringenin, and 6-geranylnaringenin, which have been determined as other most important hop flavonoids in previously reported works [71-74].

Finally, the aqueous residue collected from the extractions with cold hexane and MWCNTs/ethyl acetate, was treated with the heterogeneous system MWCNTs/chloroform, under magnetic stirring at room temperature, for 2 hours. After washing of MWCNTs with chloroform, an aliquot of the desorbed organic material was dissolved in $CDCl_3$ and subjected to the spectroscopic analysis.

The proton spectrum (figure 7) was characterized by a set of very intense resonances, due to the structures of all fatty acid chains (free and as the components of acylglycerols) present in the craft beer (up-field region between 0.70 and 1.75 ppm generated by all terminal methyl and inner- and allyl methylene protons in long acyl fatty chains, and signals at 2.28-2.34 ppm due to the bis-allyl methylene proton resonances) [75-77]. Furthermore, the sub-spectrum reported in figure 5 as the graphic enlargement of the window between 3.40 and 7.60 ppm, contained the signals attributable to choline derivatives in phosphoglycerolipids, carboxylic acids, acylglycerols, together resonances in the aromatic region that were due to not well-identified minor metabolites.

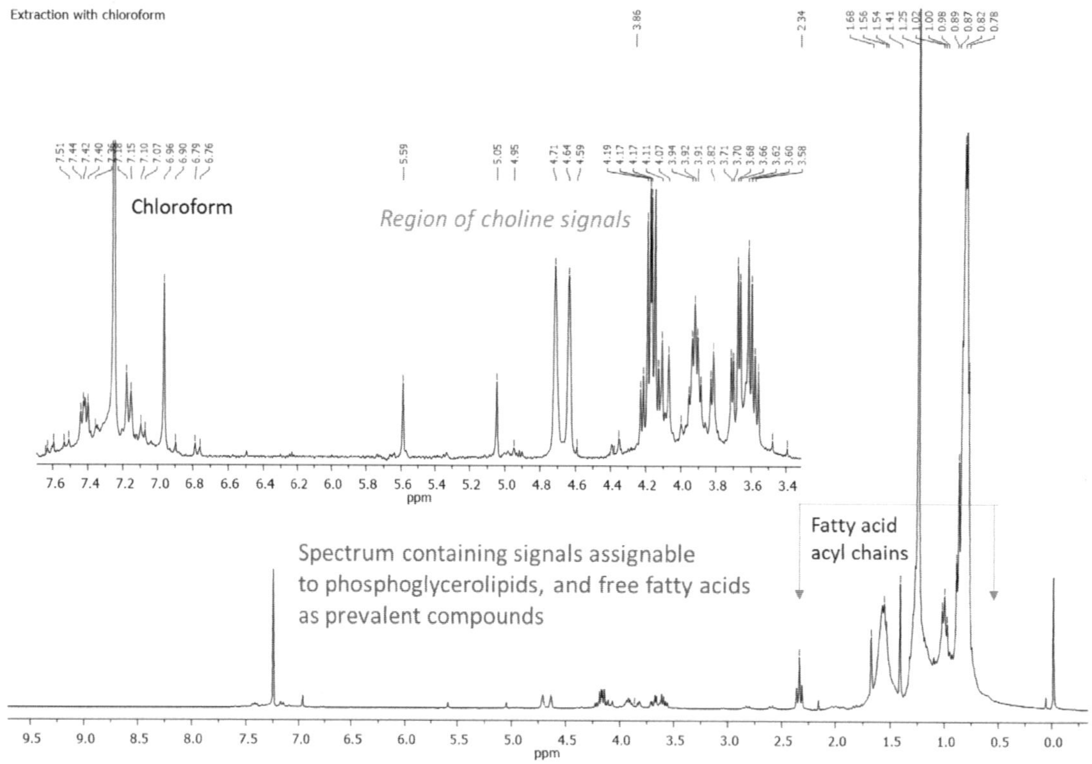

Figure 7. High-Resolution ^{1}H NMR spectrum recorded on the sample obtained by the treatment with the heterogeneous system MWCNTs/chloroform.

The aqueous phase was tested by spectroscopy for the residual contents of organic materials contained in the original sample of the craft beer dry residue (figure 8). The spectrum, recorded by applying the inverse recovery pulse sequence for the water signal suppression, and by using DMSO-d_6 for calibration, unambiguously showed the characteristics pattern of resonances attributable to the protons of the main polar chemical components of the craft beer case of study [78-80]. The analyzed sample was prevalently constituted by carbohydrates (3.00-4.00 and 4.25-6.25 ppm) obtained from hops and other raw materials used for the beer brewing, while an appreciable content in amino acids (alanine, valine, leucine, isoleucine, proline, tyrosine) was at the same time recognized by the up-field signals at 0.70-2.80 ppm, as confirmed by recording proton spectra for commercial samples of standard amino acids dissolved in water and DMSO-d_6.

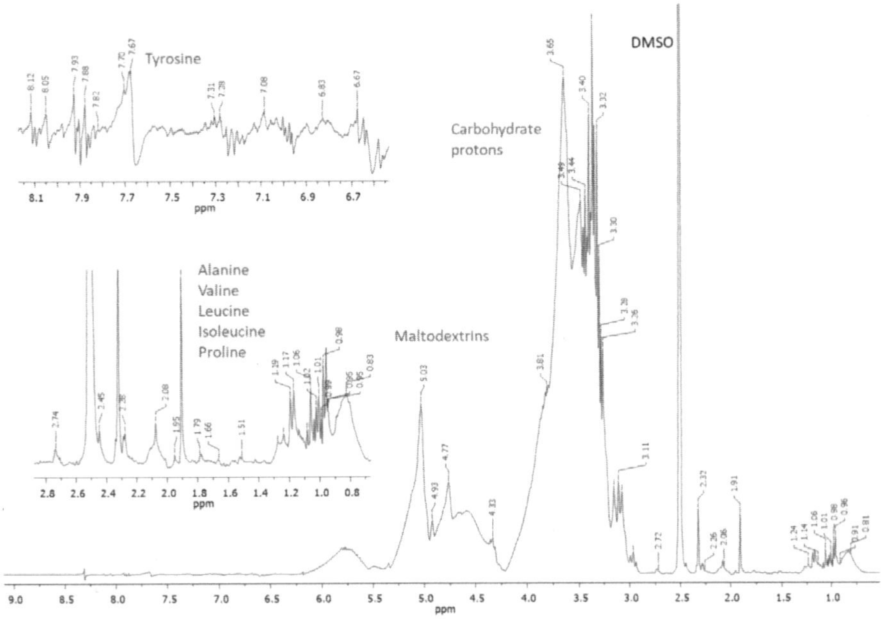

Figure 8. High-Resolution ^{1}H NMR of the residual aqueous phase recovered after the serial extraction procedure of the craft beer dry residue.

The presence of maltodextrins in the aqueous phase were confirmed by Laser Desorption Mass Spectrometry and tandem MS/MS experiments (figure 9). The spectrum showed the molecular ion ($[M+H]^+$) of these species at *m/z* 1199.63 and 1215.60, furnishing insights on the carbohydrate fraction.

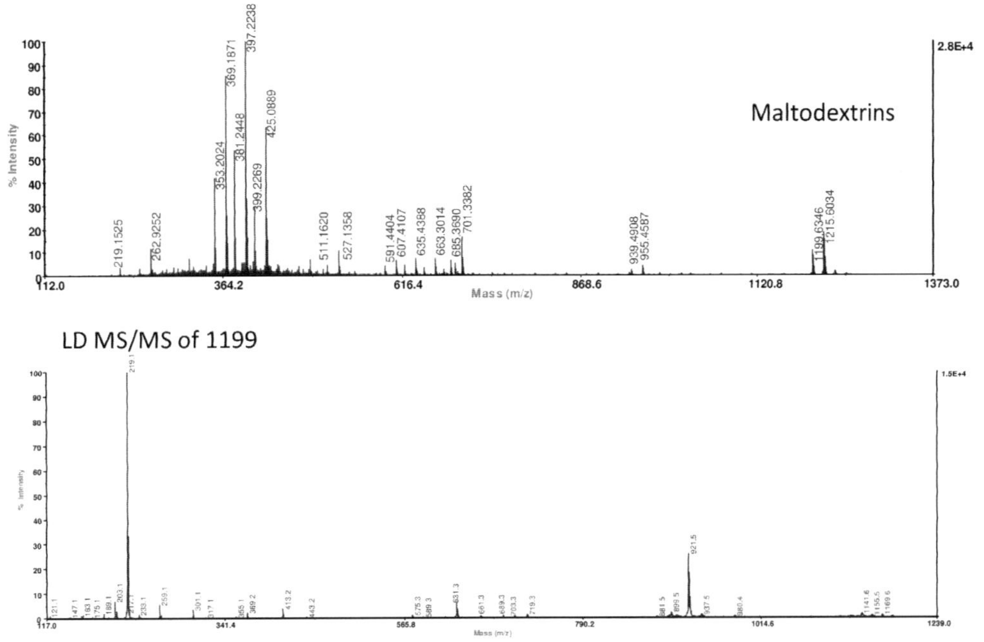

Figure 9. Laser Desorption mass spectrum, and tandem (MS/MS) mass spectrometry experiments obtained for a sample of the aqueous phase.

4. Conclusions

As far as we know, this work traces the molecular profiling of bioactive compounds of a high-quality type of craft beer produced, without additives in Calabria, a region of the Southern Italy. This paper discloses, for the first time, the use of a mixed method based on the selective solvent/multi-walled carbon nanotubes extraction and high-resolution ^{1}H NMR spectroscopy, to obtain the chemical profiles of the different classes of hop metabolites in beer. The selective extraction and sub-profile of lupulone and humulone series in hop resin, the class of antioxidant flavonoids, and free fatty acids were collectively determined, by recording a set of high resolution ^{1}H NMR spectra for samples obtained from the extraction of a small volume of craft beer with different organic solvents, and in the presence of a weighed amount of MWCNs, when required. The solvent used were of different polarities, allowing the best selective adsorption of metabolites on the surface of the carbon material. Minimal sample preparation and small volumes of beer were required. The results highlighted the highly selective separation of fractions enriched in flavonoids, polyphenols and bitter acids, affording valuable insights on the presence of the potent anticancer caffeoyl tyramine in the craft beer object of the investigation. The use of a low polar solvent, namely cold hexane, allowed the smooth separation of a fraction mainly containing the principal hop volatile components that determines the flavour of the craft beer. The residual aqueous phase predominantly contained polar compounds, namely carbohydrates and amino acids. The method might be applicable to all kinds of beer and to the soluble components of brewery processes.

References

[1] Colgrave M L, Goswami H, Howitt C A and Tanne G J 2013 *Food Res. Int.* **54** 1001
[2] Mardones W, Villarroel C A, Krogerus K, Tapia S M, Urbina K, Oporto C I, O'Donnell S, Minebois R, Nespolo R, Fischer G, Querol A, Gibson B and Cubillos F A 2020 *Microb. Biotechnol.* **13** 1012
[3] Vieira B, Klebson H, Santos S, Mendonça Barbosa A, Sawata M, da Costa Mendonça M and Ferreira Padilha F 2019 *Molecules* **24** 1568
[4] Cardoso A, Tatiana V, Pimentel C, Borges do Vale R, Santos Clementino L, Thayná E, Ferreira J, Magnani M and dos Santos Lima M 2021 *LWT* **152** 112317
[5] Hill A E and Stewart G G 2019 *Fermentation* **5** 22
[6] Van Opstaele F, De Causmaecker B, Aerts G and De Cooman L 2012 *J. Agric. Food Chem.* **60** 12270
[7] Szwajgier D 2009 J. Inst. Brewing **115** 243
[8] Nardini M, Natella F, Scaccini C and Ghiselli A 2006 *J. Nutr. Biochem.* **17** 14
[9] Oladokun O, James S, Cowley T, Dehrmann F, Smart K, Hort J and Cook D 2017 *Food Chem.* **230** 215
[10] Patzak J, Nesvadba V, Henychova A and Krofta K 2010 *Biochem. System. Ecol.* **38** 136
[11] Montanari L, Perretti G, Natella F, Guidi A and Fantozzi P 1999 *Lebensm.-Wiss. u.-Technol.* **32** 539
[12] Horn P A, Bagattoli Pedron N, Junges L H, Martinez Rebelo A, da Silva Filho H H and Bertarello Zeni A L 2021 *Eur. Food Res. Technol.* **247** 439
[13] Sastre-Serra J, Ahmiane Y, Roca P, Oliver J and Pons D G 2019 *Int. J. Food Sci. Nutr.* **70** 396
[14] Arranz S, Chiva-Blanch G, Valderas-Martínez P, Medina-Remón A, Lamuela-Raventós R M and Estruch R 2012 *Nutrients* **4** 759
[15] Aron P M and Shellhammer T H 2010 *J. Inst. Brew.* **116** 369
[16] Vanderhaegen B, Neven H, Verachtert H and Derdelinckx G 2006 *Food Chem.* **95** 357
[17] Hughes P S and Simpson W J 1993 *MBAA Techn. Quart.* **30** 146
[18] Blanco C A, Rojas A, Caballero P A, Ronda F, Gomez M and Caballero I 2006 *Trends Food Sci. Technol.* **17** 373
[19] Nord L I, Sørensen S B and Duus J Ø 2003 *Magn. Reson. Chem.* **41** 660
[20] Caballero I, Blanco C A and Porras Trends M 2012 *Food Sci. Technol.* **26** 21

[21] Intelmann D and Hoffman T 2010 *J. Agric. Food Chem.* **58** 5059
[22] da Costa Jardim C, de Souza D, Kasper Machado I C, Massochin L, Pinto N, de Souza Ramos R C and Garavaglia J 2018 *Beverages* **4** 106
[23] Hughey C A, McMinn C M and Phung J 2016 *Metabolomics* **12** 11
[24] Dusek M, Olsovská J, Krofta K, Jurková M and Mikyskă A 2014 *J. Agric. Food Chem.* **62** 7690
[25] Bullis D E 1937 *Ind. Eng. Chem. Anal. Ed.* **9** 267
[26] Franiau R and Mussche R 1974 *J. Inst. Brewing* **80** 59
[27] Mattos Duarte L, Cantarino L H, Marcone A and Leal de Oliveira A 2018 *Electrophoresis* **39** 1399
[28] Jahouh F, Saksena R, Aiello D, Napoli A, Sindona G, Kováč P and Banoub J H 2010 *J. Mass Spectrom.* **45** 1148
[29] Bartella L, Mazzotti F, Napoli A, Sindona G and Di Donna L 2018 *Anal. Bioanal. Chem.* **410** 2193
[30] Di Donna L, Taverna D, Mazzotti F, Benabdelkamel H, Attya M, Napoli A and Sindona G 2013 *Food Chem.* **141** 2328
[31] Cheiran K P, Raimundo V P, Manfroia V, Anzanello M J, Kahmann A, Rodrigues E and Frazzon 2019 *J. Food Chem.* **286** 113
[32] Zhang X, Liang X, Xiao H and Xu Q 2004 *J. Am. Soc. Mass Spectrom.* **15** 180187
[33] Kovačevič M and Kač M 2002 *Food Chem.* **77** 489
[34] Eri S, Khoo B K, Lech J and Hartman T G 2000 Journal of Agricultural and Food Chemistry **48** 1140
[35] Hrivňák, J Smogrovičová D, Nádaský P and Lakatošová J 2010 *Talanta* **83** 294
[36] Haseleu G, Intelmann D and Hofmann T 2009 *J. Agric. Food Chem.* **57** 7480
[37] Nikolić D and van Breemen R B 2013 *Curr. Anal. Chem.* **9** 71
[38] Egts H, Durben D J, Dixson J A, and Zehfus M H 2012 *J. Chem. Educ.* **89** 117
[39] Gorjanović S, Pastor F T, Vasić R, Novaković M, Simonović M, Milić S and Sužnjević D J. 2013 *Agric. Food Chem.* **61** 9089–9096
[40] Nogueira L C, Silva F, Ferreira I and Trugo L C 2005 *J. Chromatogr. A* **1065** 207
[41] Hea L, Xua Z, Hirokawa T and Shena L 2017 *J. Chromatogr. A* **1482** 109
[42] Rodda L N, Gerostamoulos D and Drummer O H 2013 *Anal. Bioanal. Chem.* **405** 9755
[43] Johnson S R, Soprano S E, Wickham L M, Fitzgerald N and Edwards J C 2017 *Beverages* **3** 1
[44] Siciliano C, Napoli A, Aiello D, De Luca P and Temperini A 2019 *IOP Conf. Ser.: Mater. Sci. Eng.* **572** 012011
[45] Siciliano C, Bartella L, Mazzotti F, Aiello D, Napoli A, De Luca P and Temperini A 2019 *IOP Conf. Ser.: Mater. Sci. Eng.* **572** 012010
[46] Siciliano C 2021 *J. Phys.: Conf. Ser.* **1960** 012006
[47] Temperini A, Aiello D, Mazzotti F, Athanassopoulos C M, De Luca P and Siciliano C 2020 *Molecules* **25** Article number 25061313
[48] Pusecker K, Albert K and Bayer E J. 1999 *J. Chromatogr. A* **836** 245
[49] Duarte I F, Godejohann M, Braumann U, Spraul M and Gil A M 2003 *J. Agric. Food Chem.* **51** 4847
[50] Rodrigues J E and Gil A M 2011 *Magn. Res. Chem.* **49 (S1)** S37
[51] Cavallini N, Savorani F, Bro R and Cocchi M 2021 *Molecules* **26** 1472
[52] Kaufmann M, Schwarz K J, Dallmann A, Kuballa T and Bergmann M 2022 *Eur. Food Res. Technol.* **248** 635
[53] Steenackers B, De Cooman L and De Vos D 2015 *Food Chem.* **172** 742
[54] Ceslová L, Holčapek M, Fidler M, Drstičková J and Lísa M 2009 J. Chromatogr. A **1216** 7249
[55] Hrnčič M K, Španinger E, Košir I J, Knez Ž and Bren U 2019 *Nutrients* **257** 1
[56] Donley J R 1992 *J. Am. Soc. Brew. Chem.* **50** 89
[57] Chen L, Zhao Q, Jin H, Zhang X, Xu Y, Yu A, Zhang H and Ding L 2010 *Talanta* **81** 692
[58] Ding J, Zhao Q, Sun L, Ding L and Ren N 2011 *J. Sep. Sci.* **34** 1463

[59] Lico D, Vuono D, Siciliano C, Nagy JB and P De Luca 2019 *J. Environ. Manage.* **237** 636
[60] De Luca P, Roberto B, Vuono D, Siciliano C and B.Nagy J 2018 *IOP Conf. Ser.: Mat. Sci. Engin.* **374** 012071
[61] De Luca P, Chiodo A, Macario A, Siciliano C and B.Nagy J 2021 *Appl. Sci.* **11** 1687
[62] De Benedetto C, Macario A, Siciliano C, B.Nagy J and De Luca P 2020 *Materials* **13** 1-14c Article number 2757
[63] De Luca P, Foglia P, Siciliano C, B.Nagy J and Macario A 2019 *Environ.* **6** 101
[64] De Luca P, Siciliano C, Macario A and B.Nagy J 2021 *Materials* **14** 2118
[65] Abdullah Z, Othmanab A and Wabaidura S M 2019 *Arabian J. Chem.* **12** 633
[66] Jeong J-H, Cho S-J and Kim Y 2017 *Bull. Korean Chem. Soc.* **38** 466
[67] Almeida C, Duarte I F, Barros A, Rodrigues J, Spraul M and Gil A M 2006 *J. Agric. Food Chem.* **54** 700
[68] Chadwick L R, Nikolic D, Burdette J E, Overk C R, Bolton J L, van Breemen R B, Fröhlich R, Fong H S, Farnsworth N R and Pauli G F 2004 *J. Nat. Prod.* **67** 2024
[69] Meneses N G T, Martins S, Teixeira J A, Mussatto S I 2013 *Sep. Purif. Technol.* **108** 152
[70] Zhu P-F, Dai Z, Wang B, Wei X, Yu H-F, Yan Z-R, Zhao X-D, Liu Y-P and Luo X-D 2017 *Nat. Prod. Bioprosp.* **7** 421
[71] Nikolić D and van Breemen R B 2013 *Curr. Anal. Chem.* **9** 71
[72] De Luca P and B.Nagy J 2020 *Materials* **13** 5508
[73] Mazza S, Aiello D, Macario A and De Luca P 2020 *Environm.* **7** Article number 37
[74] Siciliano C, Napoli A, Aiello D, De Luca P and Temperini A 2019 *IOP Conf. Ser.: Mat. Sci. Eng.* **572** 012011
[75] Napoli A, Athanassopoulos C M, Moschidis P, Aiello D, Di Donna L, Mazzotti F and Sindona G 2010 *Anal. Chem.* **82** 5552
[76] Axinte M, Vizureanu P, Nejneru C Sandu AV and Perju MC 2020 *AIP Conference Proceedings* **2291(1)** 020040.
[77] Nejneru C, Perju MC, Sandu AV, Axinte M, Quaranta M, Sandu I, Costea M and Abdullah MMAB 2016 *Revista de chimie* **67(6)** 506-511.
[78] Burduhos Nergiş D P, Nejneru C, Achiţei D C, Cimpoieşu N and Bejinariu C 2018 *IOP Conf. Ser.: Mater. Sci. Eng.* **374** 012040
[79] Baltatu MS, Vizureanu P, Cimpoesu R, Abdullah MMA and Sandu AV 2016 *Revista de chimie* **67(10)** 2100-2102.
[80] Abtew MA, Boussu F, Bruniaux P, Loghin C, Cristian I, Chen Y and Wang L 2021 *Journal of Industrial Textiles* **50(9)** 1351-1383. doi: 10.1177/1528083719862883.

Increasing the sustainability of construction sector by developing new products based on biomass and renewable polymers - bibliometric analysis

R Istoan[1,*], L D Manea[1], L Plesa[1] and M L Tintisan[1]

[1]Department of Civil Constructions and Management, Faculty of Construction, Technical University of Cluj-Napoca, 28 Memorandumului Street, 400114, Cluj-Napoca, Romania

Email: ralucafernea@ccm.utcluj.ro

Abstract. To increase the sustainability of the construction sector, one of the most important research directions in recent years has been the development of new building materials. As the construction industry is considered one of the most polluted sectors, the focus of researchers is to find new ways to shift away from traditional materials towards green materials. Those are usually derived from waste or unlimited natural-organic resources and have the role to reduce the environmental impact of the construction sector. Reaching the objective of a sustainable industry through the process of developing sustainable building materials is not about restricting the total amount of construction but paying more attention to how the new materials are designed and the traditional materials are rethought, as clean and energy efficient alternatives. The utilization of biomass in the composition of building materials present the new products as low-embodied energy materials that reduce energy use and greenhouse gas emissions. Based on this premise, the research will provide a bibliometric analysis of the studies carried out in the last years.

1. Introduction

Increasing the sustainability of the construction sector today can be achieved through several segments. One of those segments is related to building materials market. In the life cycle of a construction the first stage analyse is always related to the building materials proposed and used [1]. Worldwide consumption of building materials has increased threefold from 6.7 billion tonnes in 2000 to 17.5 billion tonnes in 2017; concrete, aggregates and bricks are the most common building materials used [2]. To reduce the negative impact of building materials on the environment, researchers have begun to analyse the trajectory of each material in detail. From the moment the raw material is extracted, to the final stage of recycling the manufactured material. Thus, a more detailed analysis of the life cycle of materials includes: extraction of raw materials, processing and manufacturing of these raw materials, transport, construction and upgrading, use and maintenance, demolition and waste management, disposal and circular processing through reuse, recycling and recovery [3]. Raw materials are essential when a new building material is designed. That is why the careful selection of raw materials plays an essential role in defining the characteristics of the final product. Biomass and renewable polymers are found in the composition of new building materials proposed by researchers to reduce the environmental impact and increase the sustainability of the sector in which these products are used [4]. According to R.A. Houghton, biomass refers to the mass of living organisms, including plants, animals, and microorganisms, or, from a biochemical perspective, cellulose, lignin, sugars, fats, and proteins, [5]. The use of biomass in the composition of new products has several advantages, such as low density, low CO_2 emissions, good thermal and acoustic insulation. On the other hand there are a number of interrogations such as the durability of biomass in the matrix of new compositions, the tendency of the biomass to water absorption, but also the negative influence on mechanical strengths [6]. Polymers come to counterbalance these effects. Polymeric materials exhibit application potentials in the building industry attributed to their excellent properties such as waterproof, anticorrosion, wear resistance, good strength, sound insulation, heat insulation [7]. The concept for new sustainable building materials is not only to use biomass and be able to use any type of polymer, but to identify and use a polymer from natural and renewable sources [8, 9].

In this direction the present paper presents a bibliometric review of the scientific papers written in the last years and published on Web of Science. The aim of the study is to show if the topic has been studied or there is a potential in the development of new composites based on biomass and renewable polymers.

2. Materials and Methods

Bibliometric review is a quantitative method for exploring and analysing large volumes of scientific data [10]. The method is used in the current research to establish if biomass and renewable polymers were studied together in obtaining new sustainable building materials.

The search criteria followed six queries, which aimed to narrow the information from general topics such as "biomass" and "polymers" to specific topic such as "hemp" and "lactic acid". The key words used were defined only as subject and the link between them will be explained below. The data based used was Web of Science and the search criteria were:

- Query 1 keywords: “biomass” and ”building material”
- Query 2 keywords: “polymer” and ”building material”
- Query 3 keywords: “renewable polymer” and ”building material”
- Query 4 keywords: “biomass” and “renewable polymer” and ”building material”
- Query 5 keywords: “hemp (as biomass)” and “polylactic (as renewable polymer)”
- Query 6 keywords: “hemp” and “polylactic acid” and ”building material”

The results obtained will be discussed and analysed in the first stage on all six queries in terms of the number of publications per year, the countries of origin of the authors and the WOS categories in which the papers were classified. In the second stage, query number four will be analysed in terms of the journals in which the papers were published, the keywords used by the authors and the countries in which the authors have an interest in this area of research. The co-word network visualization (keyword co-occurrence (1) authors keyword and (2) keyword plus) is analysed through VOSviewer.

3. Results and discussions

3.1. Analysis of the six queries

3.1.1. Annual publication trend. The number of scientific papers published in recent years on the six queries is shown in Table 1. Accordingly, the data presented summarizes the number of papers published in the last 20 years (2000-2022), the number of papers published in the last 5 years (2017-2022), and the last column shows the number of article and review papers published in the last 5 years. All the data, from Web of Science was extracted on 25 April 2022.

Based on the queries made it can be seen that the largest number of published scientific papers belong to category two, which covers polymer-based articles and building materials. The number of papers published from 2000 to date is more than 13000, while only in the last 5 years half of the total number has been published.

For the first category, the number of scientific papers reporting on biomass and building materials is 1000 over the period 2017-2022. This value, related to all papers published from 2000 to date, shows an increasing interest of researchers in the development of biomass-based building materials.

For category 3 the number of published papers is less than 500 in the last 20 years, for category 4 is 100, less than 200 for category 5 and 4 papers for the last category.

An overview of the number of scientific papers published each year from 2000 to present is given in Figure 1 for all six queries. There is an upward trend for the first two categories. If in 2000 Q1- recorded 3 works and Q2- 50 works, in 2010, the number increases at least 10 times reaching 50 Q1- works and 470- Q2- works. The 2010 values double for both categories in 2016-2017, and in 2021 Q1 shows 266 works and Q2 more than 1500.

Table 1. No. of scientific papers.

	No. of scientific papers from 2000 to 2022	No. of scientific papers from 2017 to 2022	Only articles and review articles in the last 5 years
Query 1	1638	1000	930
Query 2	13034	6590	6156
Query 3	471	307	296
Query 4	100	70	68
Query 5	170	91	83
Query 6	4	3	4

For the Q3 category, in 2012 the number of 20 published papers is reached and remains roughly constant until 2016 when it reaches 25, and in 2017 this number doubles to 50. In 2021 there will be 86 published papers. Q4 records a total of 100 scientific papers in the last 20 years, 2000-2017 values are below 10 publications/year, 2018-2020 are 12-14 papers/year, and in 2021 the number is 21/year.

Q5 shows the most significant increase in 2021 from 16 papers/year to 2022 with 25 scientific papers published in the first 4 months. Q6 with a total of 4 papers of which 2 were published in 2022.

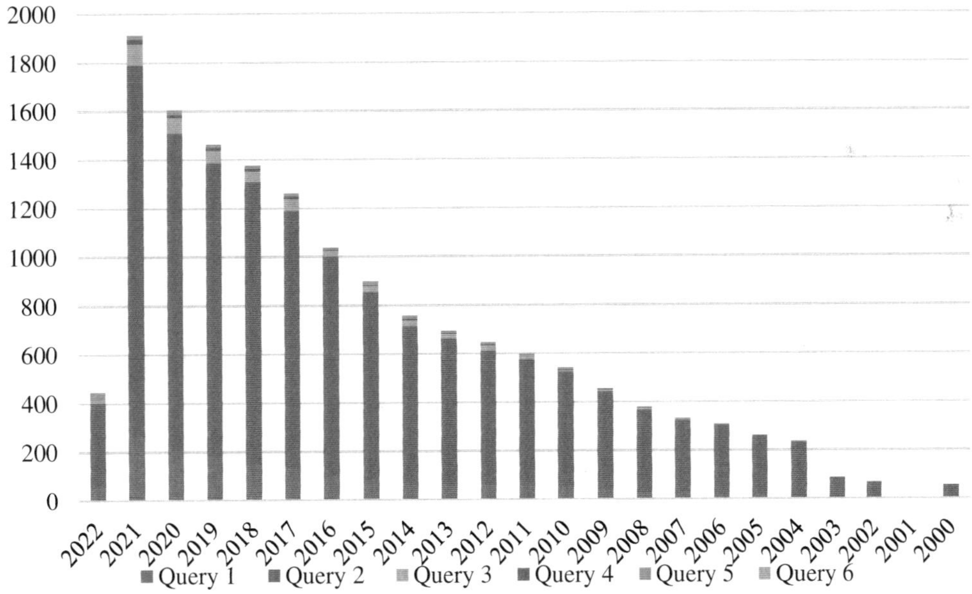

Figure 1. Annual publication trend.

The situation of scientific works in the last 5 years is shown in Figure 2. It can be seen that Q1 increases on average by 20 papers each year from 2017- 2020, and by 50 papers 2020-2021. Q2 shows the same trend, 100 papers/year from 2017-2020, and the value doubles from 2020-2021. For Q3 the values vary between 44 and 86, for Q4 between 11-21 and for Q5 between 9 and 25.

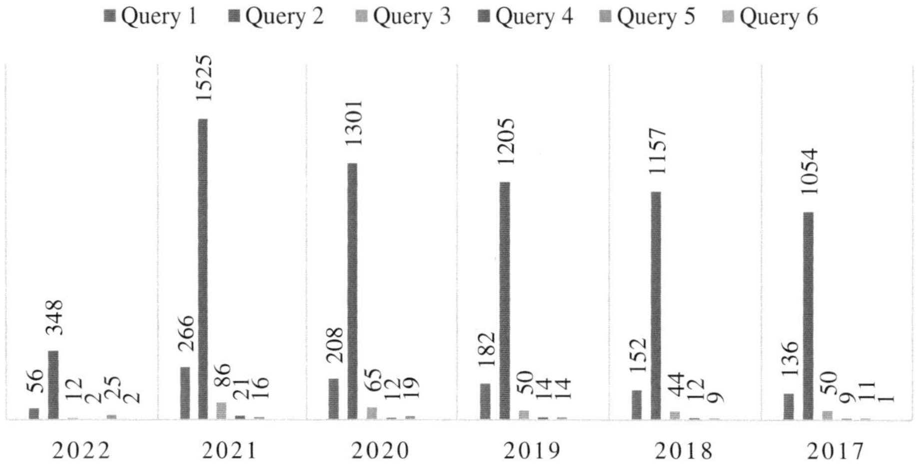

Figure 2. Scientific papers published in the last 5 years.

3.1.2. Web of Science Categories. In terms of the main WoS categories in which papers have been published relating to the six queries, it is interesting to note that the top 5 categories, presented for each query in descending order, are repeated in all categories. In the overall analysis, Materials Science, Polymers, Green Sustainable Science, Chemistry, Engineering, etc. are included, Table 2.

Table 2. Main categories in WOS.

Query 1	Query 2	Query 3	Query 4	Query 5	Query 6
Energy Fuels	Materials Science Multidisciplinary	Polymer Science	Chemistry Multidisciplinary	Materials Science Composites	Materials Science Composites
Environmental Sciences	Chemistry Multidisciplinary	Chemistry Multidisciplinary	Green Sustainable Science Technology	Polymer Science	Chemistry Physical
Engineering Chemical	Polymer Science	Materials Science Multidisciplinary	Polymer Science	Materials Science Multidisciplinary	Engineering Manufacturing
Green Sustainable Science Technology	Chemistry Physical	Green Sustainable Science Technology	Materials Science Multidisciplinary	Engineering Manufacturing	Materials Science Multidisciplinary
Chemistry Multidisciplinary	Physics Applied	Energy Fuels	Biotechnology Applied Microbiology	Engineering Environmental	Materials Science Textiles

In the graph presented in Figure 3 five categories were selected: Civil Engineering, Construction Building Technology, Materials Science, Green Sustainable Science and Material Science Paper Wood. These were found in almost all the queries and their specificity is close to the main idea of the research, which is the development of sustainable materials based on biomass and renewable polymers. The values in the graph are expressed as a percentage of the number of papers analysed from 2000 to date.

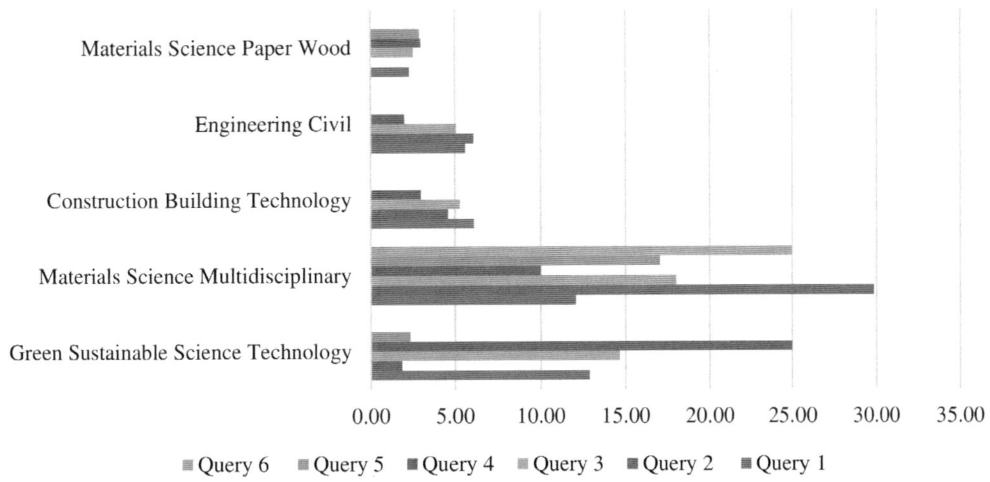

Figure 3. WoS categories related to construction/building research area.

Q1- Biomass and Building Materials is covered 12% in Green Sustainable/Materials Science, 6% in Construction/Engineering Civil and 2% in Materials Science Paper Wood. Q2 and Q3- polymers and building materials, respectively renewable polymers and building materials are found in 5% of all papers published in the Construction Building Technology and Engineering Civil categories. Q4 and Q5, in total 35% and 20% respectively in the categories Green Sustainable Science Technology and Materials Science Multidisciplinary. Q6 -(hemp and polylactic acid and building materials) is 25% in Materials Science Multidisciplinary.

3.1.3. Publication data by country. The countries where interest in the development of sustainable biomass and/or polymer-based building materials is growing are shown in Table 3. Thus, for each query the top four countries are presented in descending order of the number of published articles. As a general observation it can be mentioned that the USA and China are the first two ranked in most of the proposed queries. If we look at the situation in Europe: France, Germany, Spain, and Italy are in the top positions in the analysis.

Table 3. Main countries.

Query 1	Query 2	Query 3	Query 4	Query 5	Query 6
China	China	USA	USA	Italy	Canada
USA	USA	China	China	China	India
Spain	Germany	Germany	France	Spain	New Zealand
Germany	France	France	Germany	India	Spain

The analysis shown in Table 4 is also interesting, capturing the percentage share of the first two areas regarding the number of papers published from 2000 to date. Thus, out of the total number of China and US works, it incorporates 35% on query 1, and about 50% on query 2. For query 3 (renewable polymers and building materials) the US and CHINA together account for 36% and 42% of Q4. On Q5 we have 30% Italy and China, and on Q6 we have 50% Canada and India.

Table 4. Papers percentage of the first two countries.

Query 1		Query 2		Query 3		Query 4		Query 5		Query 6	
China	19.96	China	25.88	USA	18.89	USA	24	Italy	16.47	Canada	25
USA	14.89	USA	21.74	China	17.19	China	18	China	12.94	India	25
	34.86		47.63		36.09		42		29.41		50

3.2. Analysis of the Query 4

Because the paper aims at analysing mainly query 4, which will be further developed in the laboratory, figures 4 to 6 have been made and will be analysed only for this situation. Question 4 is based on keywords related to biomass, renewable polymer and building materials. A total of 100 papers have been found from 2000 to date, of which 70 papers have been published in the last 5 years and 20 scientific papers only in 2021.

3.2.1. Publication data by country. Figure 4 shows the distribution of published papers by area. Thus, the darker colored areas represent the publication of a higher number of scientific papers on the topic developed in Q4 compared to the lighter-colored areas, where the number of papers is reduced. It can be seen that, after the USA and China, France, Germany, the Netherlands, Brazil, Belgium, England, Japan, Portugal, Russia, etc. are the countries where articles have been published in recent years. For Romania, the number of published papers found in this query was two.

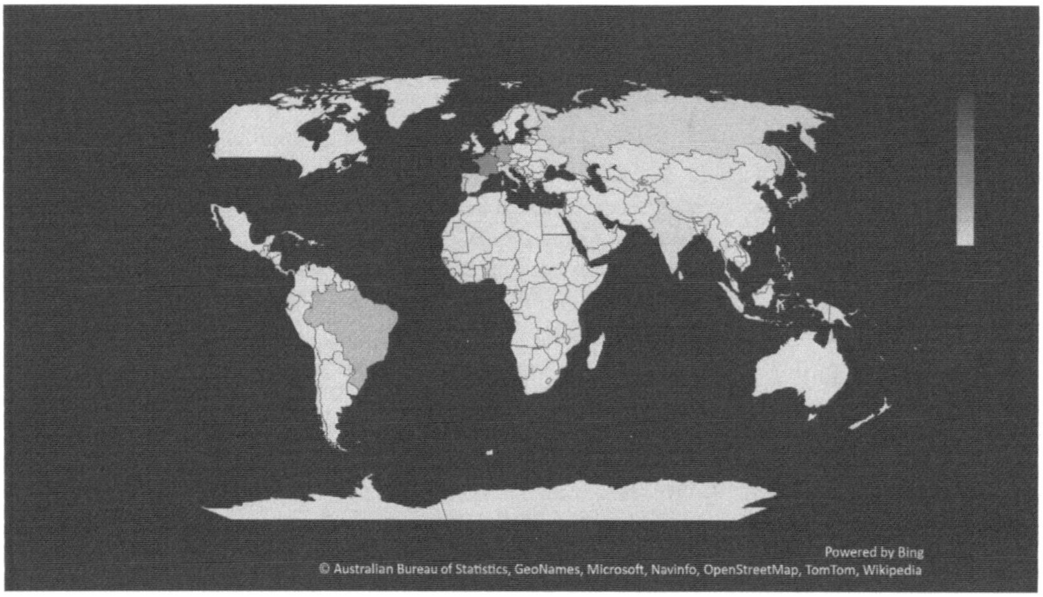

Figure 4. Query 4 in countries.

3.2.2. Publication data by keywords. Figures number 5 and 6 show the situation of the keywords chosen by the authors, as well as the extra words. Author Keywords, those provided by the original authors, and Keywords Plus, generated by an automatic computer algorithm, are words or phrases that appear frequently in the titles of an article's references and not necessarily in the title of the article or as Author Keywords [11]. For both situations that have been rendered by figure 5 and 6, an occurrence of two has been chosen [12]. It is important to note that the keywords used are highlighted according to the year of publication of the articles [13,14].

Figure 5 shows that the main words chosen by the authors, such as biomass and lactic acid, predominate from the very beginning of the published articles, whereas the key words related to the "green building" segment do not appear until 2020.

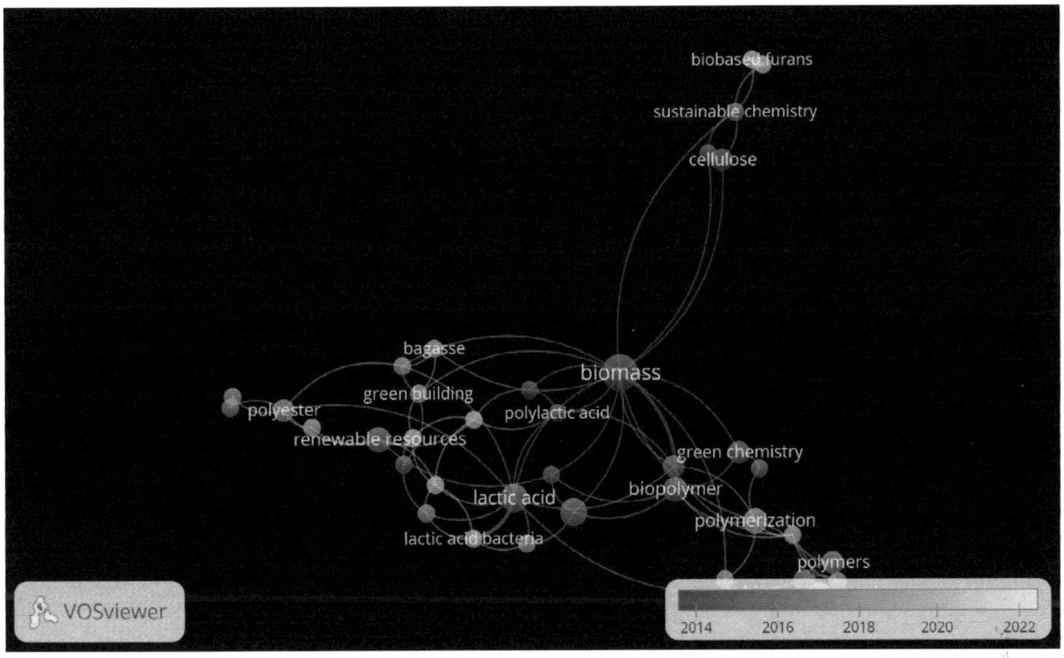

Figure 5. Query 4 – Authors Keywords.

A more complex analysis of the keywords found in published articles is shown in Figure 6. Keywords plus finds 115 important words at an occurrence of two. The numbers of keywords related to the building segment, such as building blocks, mortar, mechanical properties, performance, high-performance, behaviour, life-cycle assessment, increased especially from 2018 to present.

Based on the information presented in the bibliometric review, it can be concluded that there is an upward trend towards the use of biomass and renewable polymers, especially lactic acid, in the construction sector, in the matrix of final products [15-17].

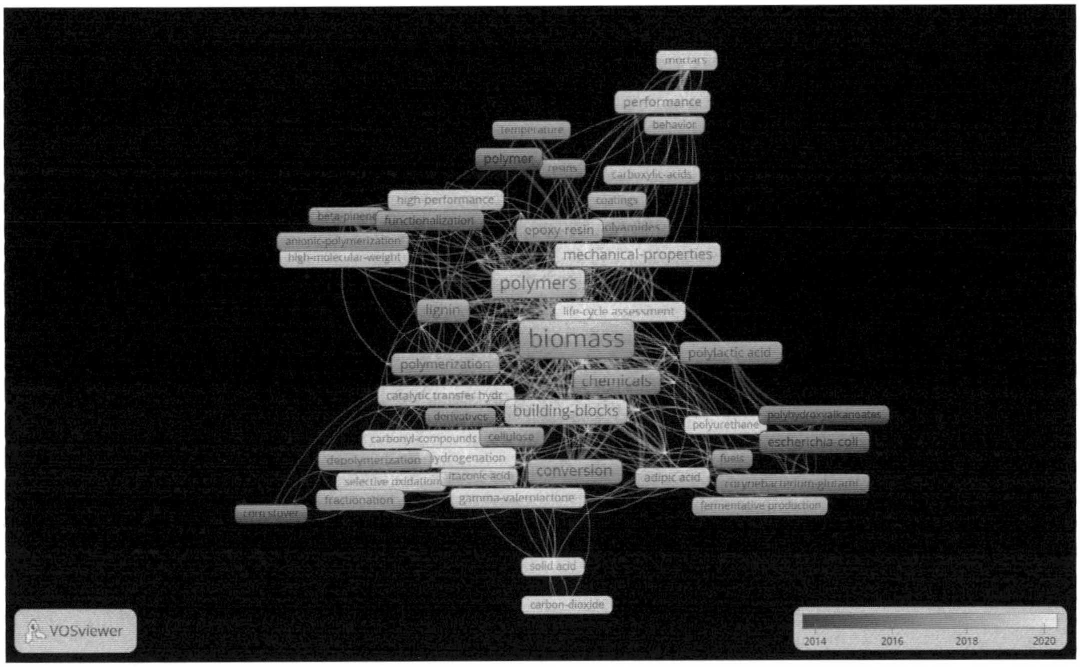

Figure 6. Query 4 – Keywords Plus.

4. Conclusions

The bibliometric analysis allowed an in-depth study of the number of papers that refer to the use of biomass and polymers in the building materials market. Throughout the study, the increasing interest of researchers towards the use of renewable and unlimited natural resources in the composition of building materials was observed. As far as renewable polymers are concerned, they are also becoming increasingly studied in this segment of the construction industry. The methodology used allowed the following conclusions to be drawn:

- the highest numbers of papers published from the six queries addressed was related to building materials and polymers over 10000 in the last 20 years
- the number of papers published on "biomass and building materials" is double that of "renewable polymers and building materials"
- in the last five years there has been an increasing trend towards the study of building materials incorporating renewable natural resources
- the main category in WoS in which papers on "biomass and building materials" have been published is energy fuels, therefore there is space to further research of biomass as a component in the matrix of composite materials
- the main countries around the world showing increased interest in the study of friendly composite materials are the USA and China, and in Europe: Spain, Germany, Italy and France.

The query "biomass" and "renewable polymer" and "building material" was analysed in more detail in terms of the countries in which the articles were published and their keywords and it was observed:

- the linking of words related to the building materials segment with biomass and regenerable polymers has been emerging since 2020, which denotes that the topic is quite early as a research topic and has potential for future research.
- The country that has published the most studies on the topic of biomass and renewable polymers in the construction segment is the USA, which actually shows increasing interest in the development of new composite materials based on natural resources.

Based on all the above, there is potential in the development of building materials that meet today's needs and perform as well or better than conventional materials. The current study reinforces the idea to continue the research with the development and production in the laboratory of a construction material, in the form of plaster, incorporating biomass (hemp wood waste) and renewable polymer (lactic acid).

Acknowledgments

This paper was financially supported by the Project "Entrepreneurial competences and excellence research in doctoral and postdoctoral programs - ANTREDOC", project co-funded by the European Social Fund financing agreement no. 56437/24.07.2019

References

[1] Carabaño R, Hernando S M, Ruiz D and Bedoya C 2020 *Revista De La Construcción* **16**(1) 22–33
[2] Huang B, Gao X, Xu X, Song J, Geng Y, Sarkis J, Fishman T, Kua H and Nakatani J 2020 *One Earth* **3**(5) 564–573
[3] Najjar M, Figueiredo K, Palumbo M and Haddad A 2017 *Journal of Building Engineering* **14** 115–126
[4] Christelle R, Sary A, Audrey V, Delphine R and Yves A 2022 *Renewable and Sustainable Energy Reviews* **156** 111962
[5] Houghton R A 2008 *Encyclopedia of Ecology* pp 448–453
[6] Vo L T T and Navard P 2016 *Construction and Building Materials* **121** 161–176
[7] Shen J, Liang J, Lin X, Lin H, Yu J and Yang Z 2020 *International Journal of Polymer Science* **15** 8838160
[8] Al-Jahwari F S and Pervez T 2020 *Encyclopedia of Renewable and Sustainable Materials* **5** 200–206

[9] Pellis A, Malinconico M, Guarneri A and Gardossi L 2021 *New Biotechnology* **60** 146–158
[10] Donthu N, Kumar S, Mukherjee D, Pandey N and Lim W M 2021 *Journal of Business Research* **133** 285–296
[11] Zhang J, Yu Q, Zheng F, Long C, Lu Z and Duan Z 2016 *J. Assn. Inf. Sci. Tec.* **67** 967–972
[12] Ahmad W, Ahmad A, Ostrowski K A, Aslam F and Joyklad P 2021 *Case Studies in Construction Materials* **15** e00683
[13] Nergis D P B, Cimpoesu N, Vizureanu P, Baciu C and Bejinariu C 2019 *Materials Today: Proceedings* **19** 969–978
[14] Cimpoesu N, Stanciu S, Vizureanu P, Cimpoesu R, Achitei DC and Ionita I 2014 Journal of mining and metallurgy section B-metallurgy **50(1)** 69-76. DOI: 10.2298/JMMB121206010C.
[15] Sioustis IA, Axinte M, Prelipceanu M, Martu A, Kappenberg-Nitescu DC, Teslaru S, Luchian I, Solomon SM, Cimpoesu N and Martu S 2021 *Applied Sciences* **11(9)** 3824
[16] Perju MC, Nejneru C, Vizureanu P and Axinte M 2013 *Metalurgia International* **18** 174-177.
[17] Tadesse MG, Nagy L, Nierstrasz V, Loghin C, Chen Y and Wang L 2018 *Materials* **11(12)** 2466. doi: 10.3390/ma11122466.

Study on behaviour of electrical insulating materials combined with conductive elements

G Buică[1], A E Antonov[1], C Beiu[1], D Pasculescu[2] and C Sipos[3]

[1] National Research and Development Institute for Labor Protection "Al. Darabont ", Bucharest, Blvd. Ghencea no. 35A, sect. 6, Romania
[2] University of Petrosani, 20 Universitatea st., Petrosani 332006, Hunedoara, Romania
[3] SC Nakita Prod Comimpex SRL, 150, December 1, 1989 st., Targu Mures, Romania

Email: gbuica@protectiamuncii.ro

Abstract. The study presents the research results on the identification of technical and safety requirements specific to work equipment made of electrical insulation materials combined with conductive elements, to guarantee safety in use. The identification of the technical and safety requirements specific to the work equipment that is provided with electrically insulating materials combined with conductive elements was carried out, taking into account the safety conditions that must be guaranteed from the design and manufacturing phase. The national technical norms for live work stipulate that the work equipment used for electrical installations must be certified from the point of view of safety and health at work. Research study shows the behavior of various electrical insulation materials combined with conductive elements under the influence of mechanical, electrical, thermal and environmental factors to verify and certify the safety function that must be guaranteed in use. The research results consist of the development of the experimentation and evaluation methodology concerning the legal regulations in force and the evaluation of compliance with the technical and safety requirements identified for work equipment that is provided with electrical insulation materials combined with conductive elements - occupational health of workers and a safe work environment.

1. Introduction

Efforts at the European and national levels to maintain a safe and healthy workplace have resulted in a decrease in workplace accidents and occupational diseases, with many major and small businesses choosing zero accidents as their aim for workplace safety and health.

It is well known that risk workplace assessments enable risk factors to be eliminated by providing adequate working conditions, limited risk exposure, proper personal safety equipment, medical surveys, and ergonomic workplace design, and that when risks cannot be eliminated, they must be reduced to the level of residual risks, which must be adequately controlled to ensure healthy workplaces for all workers [1-3].

Ensuring the integrity and health of workers also requires guaranteeing safe work equipment (EM), which involves an assessment and compliance with their technical and occupational safety and health requirements from the design and execution phase.

Thus, in addition to the technical, ergonomic and reliability requirements, work equipment must guarantee an adequate level of safety, to ensure compliance with the technical and occupational safety and health requirements to ensure the safety at work of workers.

Studies, technical diagnoses, as well as evaluations and technical expertise performed on work equipment in use have led to the identification of potential sources of risk such as:

- non-correlation of the specific operating conditions of the work equipment with the technical conditions and the safety requirements for materials;
- the use of materials that do not comply with the technical requirements in the execution and assembly stages;
- failure to identify / disregard risk factors associated with operating and location conditions;
- Inadequate evaluation of WE's technical-functional characteristics in relation to the technical work system's performance requirements.

The paper presents some of the findings from the project's research on work equipment made of electrical insulation materials combined with conductive elements: (PROC)", POC P 40 182 / 111954, ID / SMIS Code 2014+: P 40 182 / 111954.

Studies were conducted on two types of work equipment, alternating current and direct current electric vehicle charging stations, in a subsidiary project, and studies were conducted on other categories of equipment used in electrical installations, including testing laboratories for materials and protective equipment against electric shock, in another subsidiary research project.

Charging / recharging stations are used to charge electric vehicles (VE) with alternating or direct current, and provide all of the necessary functions for the conversion of voltage and / or electric current supplied by the electricity network or grid [4].

Voltage indicators are portable protective devices used to detect or signal / indicate clearly the presence or absence of voltage at rated frequencies [5].

In addition to functional, technical, and ergonomic requirements, the two types of work equipment must provide an adequate level of safety in order to comply with technical and occupational safety and health requirements [6-8].

Following the need to ensure compliance with the essential health and safety requirements, defined as safety objectives in Annex No. 1 of Directive 2014/35/EU, in conjunction with the safety and health requirements set out in Annex No. 1 of Directive 2009/104/EC, the research studies in the two subsidiary projects focus primarily on the analysis carried out in order to identify the measures needed to increase the rate of compliance in the use of the directives.

The methods used are:

- identification of professional risks applicable to charging stations for electric vehicles, respectively to voltage indicators that fall into the category of electrical equipment intended for use within certain voltage limits, and which is the subject of analysis in the project;

- identification of professional risks applicable to charging stations for charging electric vehicles, respectively to voltage indicators that fall into the category of electrical equipment intended for use within certain voltage limits, and which is the subject of analysis in the project;

- determining the safety requirements for charging stations for electric vehicles, as well as voltage indicators, in accordance with the legal provisions set forth in Annex no. 1 of GD no. 409/2016, which adopts the LVD Directive 2014/35/EU (electrical equipment design directive intended for use within certain voltage limits);

- establishing the safety requirements applicable to charging stations for charging electric vehicles, respectively to the voltage indicators according to the legal provisions established by the safety standards (for electric vehicle charging stations SR EN 61851-1:2019, respectively SR EN 61243-3:2015) specific analyzed within the subsidiary projects [4,6-8].

This study describes the findings of a study focused solely on the housings of these equipment.

2. Identification of the specific technical and thermal safety requirements that must be met by the housings made of electrical insulating materials, combined with conductive elements, which are part of the work equipment

The analysis of applicable European and national law, as well as national technical and occupational safety and health regulations, as well as national standards and codes of good practice, served as a starting point for identifying technical and occupational safety requirements.

When creating technical and occupational safety criteria, the users' circumstances for ensuring worker safety and health, as well as technology needs, were taken into consideration.

Electrical equipment (electric vehicle charging stations) and protective equipment (voltage indicators) are two types of work equipment that must be designed and manufactured in such a way that, once assembled, they assume no danger to users if used in acceptable working conditions and according to the instructions for use. [4,7].

This general requirement also applies to housings, attached components, work equipment accessories.

The importance of housings in ensuring protection against electric shock by direct contact was taken into consideration while developing specific technical and occupational safety criteria. The housings must meet the following criteria in this regard:

- dimensional and ergonomic features
- dielectric strength at normal temperature, as the case may be, dielectric strength at extreme temperatures (requirement for equipment mounted in outdoor environments or used in extreme environmental conditions);
- resistance to mechanical shocks;
- resistance to thermal shocks;
- resistance to climate stress;
- resistance to chemical stress.
- endurance and aging resistance [4,9].

The carcass components' materials must be chosen to withstand the electrical, mechanical, thermal, and chemical stress that they will face during usage [7,9].

These materials must have adequate resistance to aging and must not spread the flame.

In the market, users have requested additional conditions to ensure the equipment housings addition to robustness, impact resistance, shock resistant, weather-resistant, so that over time the housings do not fade - colourfast (colour resistance of a material to fading or leaking of paint, in time) [4].

The technical and work safety requirements applicable to the carcasses and their materials have been experimentally certified by tests.

3. Experimental procedures

The studies on the behaviour of the carcasses were performed on three models of charging stations for electric vehicle recharging stations made of polycarbonate, PVC combined with aluminium elements and polycarbonate combined with fibreglass, respectively three models of voltage indicator housings made of PVC, polycarbonate.

The tests were performed on 3 specimens of each model.

Due to the non-conformities registered, the carcasses underwent changes during the study, both in material and in design.

The housings of two types of equipment were analysed, diagnosed, and evaluated: rechargeable electric car charging stations and voltage indicators in use.

The appropriate test techniques for evaluating the safety quality of electric car charging stations and voltage indicators must be found and implemented. In an ideal world, test techniques would perfectly duplicate the conditions that occur in practice; however, this is not achievable, thus testing on these hazards are conducted by replicating as closely as possible the real-world impacts [10].

Laboratory tests were performed on the housings of charging stations for electric vehicles.

The voltage indicators were subjected to two types of tests:

- Laboratory tests on technical and functional characteristics
- Tests in real operation conditions (by checking the operation of the voltage indicator in LV electrical installations - power station) [10].

Enclosures with all components (buttons, screens), as well as enclosures taken from the production lines of some manufacturers with electrical safety products / equipment, were tested.

The tests were carried out in accordance with the verification techniques described in the appropriate Romanian standards, the procedures of the ICSPM - CS certification body, and the INCDPM testing laboratories for "Electrical and Mechanical Risks" (REM) and “Personal Protective Equipment” [11].

Mechanical, climatic, and electrical tests were carried out in accordance with the logical scheme of application of the procedure for evaluating and diagnosing the safety quality of equipment and means of protection established at the REM Laboratory and the Certification Body - Work Equipment.

3.1. Mechanical shock check

Verification of the degree of protection provided by the equipment casings against external mechanical shocks (e.g. protection against impact, vandalism) was carried out on:

- Fully equipped housings conditioned for 24 hours at ambient laboratory temperature;
- Fully equipped housings after heating test;
- Housings (empty) conditioned at -25°C for 60 hours;
- Housings (empty) conditioned at +60°C for 60 hours.

Shock checks were performed immediately after each conditioning.

After each conditioning session, shock tests were conducted.

The test for heating the housings consisted of locating and measuring the hottest point on the housing of fully completed equipment that was supplied with a voltage of 1.2 times the rated voltage in alternating current and at the manufacturer's maximum degree of use. Measurement of the warmest spot on the equipment housing after it has been disconnected or has lost power. The test was deemed appropriate if the heating of the housings did not exceed 50 K, and the heating of the parts where users had access / placed the hands (e.g. handles) did not exceed 30 K [4].

There were no hot areas above 50K or 30K in any of the 9 samples analyzed.

The mechanical strength test IK was performed immediately after each conditioning, correspondingly immediately after verifying the heating of the equipment during operation, in accordance with the standard SR EN 62262: 2004.

The mechanical shock resistance test was performed using the pendulum method.

A hammer is attached to the end of an oscillating arm, which rotates around a horizontal axis to create the pendulum. The hammer moves vertically due to gravity [4,13].

The 1.5 kg hammer, which is fixed at the bottom end and has an oscillation radius of 1 m, produces a shock of 6J with a drop height of 400 mm and can be replaced depending on the value of the shock imparted.

The equipment housings were tested in accordance with the manufacturers' technical specifications as well as the appropriate standards.

According to SR EN 61851-1: 2019 [4, 14], the outside casing/enclosure of VE mode 2 charging stations must have a minimum degree of protection against mechanical impacts of IK 08 [4,14].

The shock administered to the housings of the electric vehicle charging stations under investigation was equivalent to the manufacturer's IK 10 index, which is equivalent to a shock of 20 J. A 5 kg hammer was dropped from a height of 40 cm to produce a 20 J shock [4,15].

The specimens were mounted on a rigid surface with the point of contact for each shock being the intersection of the hammer’s trajectory with the vertical plane running through the oscillation axis.

Each specimen was shocked perpendicular to the surface and at two spots that were likely to be sensitive or vulnerable during use.

If the specimens have not suffered any visible cracks, damage, or destruction and have preserved their original qualities, the testing is declared satisfactory.

One test specimen (PVC test tube code anonymous RE 0859 -3) had cracks in the area of the light signals / display, out of nine (enclosures of electric car charging stations).

After verifying for mechanical shock resistance, the specimens were tested for dielectric strength, not only to ensure that the electrical quality was safe, but also to rule out the presence of faults / concealed / unnoticed cracks caused by the shock tests. Because the housings are components of low-voltage equipment, the specialists selected to test them for 60 seconds at a 5250 V alternating voltage and a 50 Hz industrial frequency. The leakage current value must not be greater than 1 mA.

The specimen with the RE 0859 -2 anonymous code has been punctured. Cracks in the corner area were discovered after evaluating the enclosure, forcing the producer to modify the PVC case and merge it with aluminium pieces.

Figures 1 a) and 1 b) show the results of dielectric strength testing of electric car charging stations built of PVC, which were conditioned at room temperature for 24 hours before being subjected to mechanical shock testing.

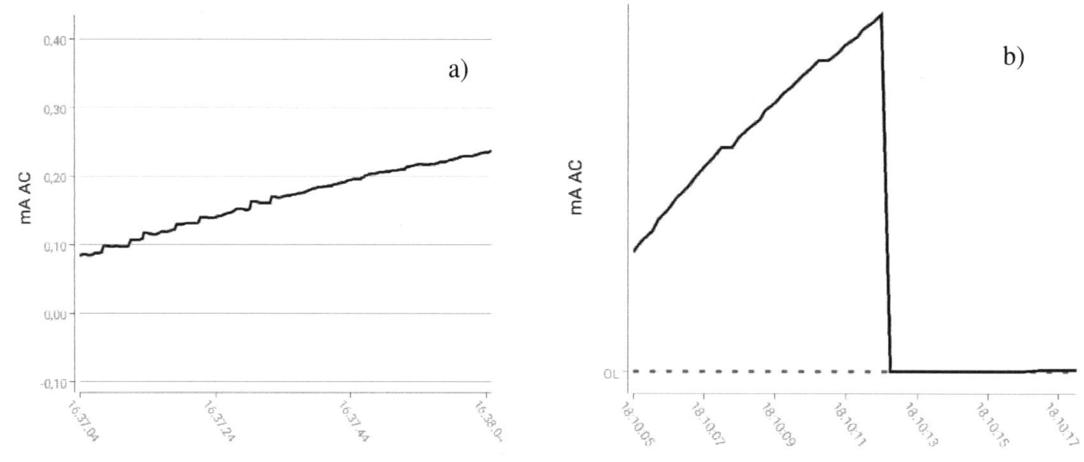

Figure 1. a) After conditioning for 24 hours at ambient temperature and shock testing, leakage current values - test sample anonymous code RE 0859-1. b) Current leakage values after 24 hours of conditioning at ambient temperature and shock testing - anonymous code RE 0859-2.

The same approach was used to evaluate the modified housings in conjunction with aluminum elements. After conditioning at ambient temperature, high temperatures, and a heating test, the housings' robustness was assessed. Testing the dielectric strength confirmed the presence of cracks or unobservable damage during the visual assessment. There were no data recorded that exceeded the maximum leakage current allowed. Due to the humidity, there were discharges on the surfaces of the specimens conditioned at severe temperatures.

The results of dielectric strength testing of recharged electric vehicle charging station housings, conditioned at severe temperatures and shock tested, are shown in Figures 2 a) and 2 b).

The housings of electric vehicle charging stations made of polycarbonate combined with fibreglass showed no cracks or damage after being subjected to a 20 J shock and did not exceed the permissible leakage current limit, demonstrating the increased degree of shock protection and insulation properties.

The technical and safety aspects of low voltage indications are assessed in the second subsidiary project, using the same experimental and assessment technique as for electric vehicle charging stations. The degree of protection provided by the equipment housings against external mechanical shocks is verified immediately after the housings have been exposed to the environment, severe temperatures, and the heating test has been completed.

The shock resistance of voltage indicator housings was tested at 6 J, as per the manufacturer's specifications and the standard SR EN 61243-3: 2015.

Two PVC test pieces with the codes RE 0860 -2 and RE 0859 -3 found visible cracks in the display area and the joint between the display and the case following the impact. The two specimens were tested with the test voltage at piercing to verify the degree of impact damage.

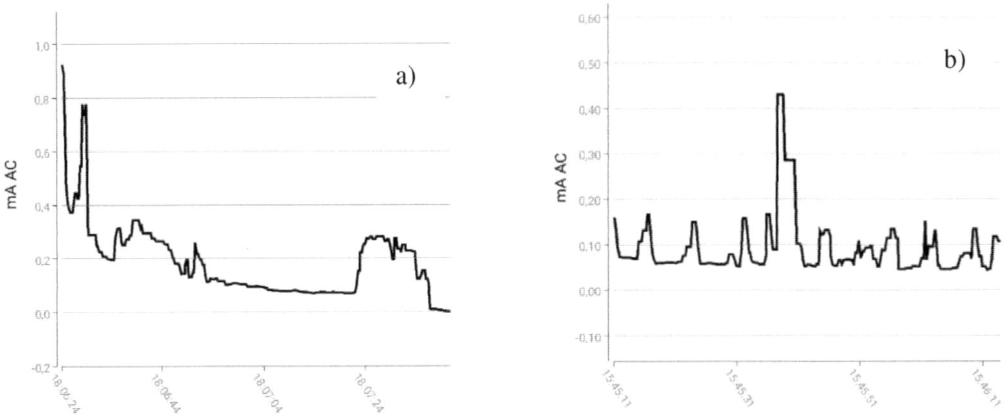

Figure 2. a) Leakage current readings - after conditioning at + 60 ^{0}C and shock testing with RE 0859 -11 anonymous code. b) After conditioning at – 25 ^{0}C and shock testing, leakage current values - test tube anonymous code RE 0859 -12.

Voltage indicators that fall within the category of safety equipment for use in low voltage installations were tested for 60 seconds at an applied AC voltage of 5250 V, at an industrial frequency of 50 Hz, with the leakage current not exceeding 1 mA.

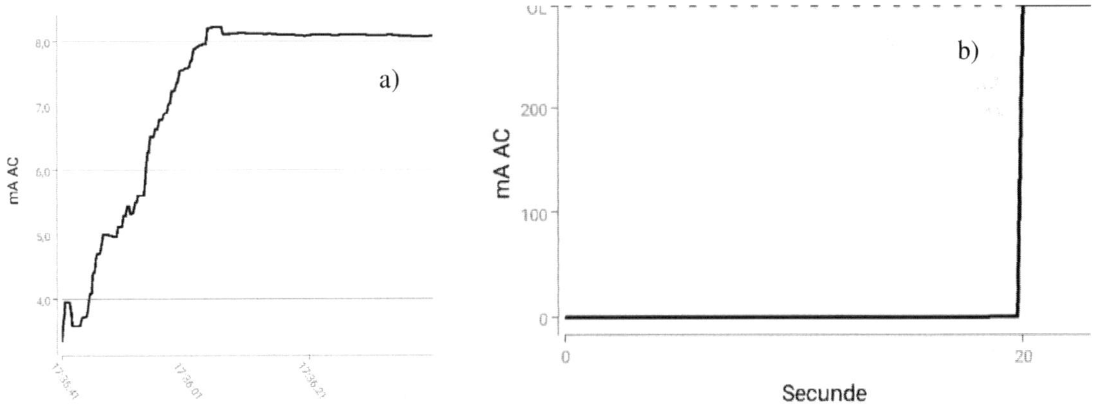

Figure 3. a) Leakage current values - after the shock test RE 0860 -2 anonymous code. b) Leakage current values - after the shock test RE 0860 -3 anonymous code.

Figure 3 a) and Figure 3 b) show the results of the voltage indicator housings' dielectric strength test following the shock test. The specimen marked RE 0860 -2 registered leakage currents more than 1 mA, while the specimen labeled RE 0860 -3 was perforated in the impact zone.

The specimens passed the impact resistance and dielectric strength tests after the producers adopted various connection solutions between the display housing and the indication housing. There were no readings above the permissible limit of 1 mA.

Cracks were discovered in the polycarbonate test specimens in the area of the battery compartment during the verification of the dielectric strength, which led to the redesign of the housing model, the manufacturer choosing a different battery model, and changing the design and dimensions of the battery compartment.

Following the mechanical shock test of the fully equipped equipment housings, functional testing in the laboratory and LV electrical installations and extremely low voltage were conducted to ensure that they retained their original characteristics.

Electrical and electronic components' operation can affect the mechanical, physical, and insulating properties of housings, such that a material with high dielectric losses will heat up more than one with low dielectric losses [10].

4. Results and discussions

The findings of research studies conducted on the carcasses of work equipment, particularly protective equipment used in electrical installations, have highlighted the importance of verifying the results of destructive mechanical tests not only visually but also by testing the material's dielectric strength.

In the event of an external mechanical shock, the electrical insulating materials of protective equipment, particularly the protective equipment held in hand by voltage indicators, must keep their characteristics and technical properties [16].

Insulation resistance measurement is one of the required periodic tests to assure the safety of employees and electrical installations. The purpose of these tests is to prevent mishaps caused by electric shock and equipment damage [17].

The information collected through the application of the assessment technique can be used to verify that the goal of safe work equipment is realized, as well as to strike a balance between technical and safety criteria, as well as reliability and cost considerations [18].

The implementation of the assessment technique validates that the technical functions' requirements are met while keeping the operational indicators' values within the prescribed limitations [19-23].

The research emphasized the importance of developing a technique for analyzing occupational risks in order to avert hazards that can lead to disasters, particularly in the energy industry, through a decision support system for users of work equipment.

5. Conclusion and perspectives

The project's studies led to the development of a methodology for evaluating work equipment that is provided with electrical insulation materials and material electrical insulation combined with conductive elements, with the goal of reducing the risks associated with not correlating the specific operating conditions of work equipment with the technical conditions and safety requirements for materials.

The studies will be completed with the development of new prototypes of work equipment for the protection of workers in electrical installations and a decision support system complete with a training module specific to identified occupational risks, taking into account the objectives of the research being implemented in the project. as a beneficiary enterprise, at the level of a testing laboratory of a manufacturer of work equipment and means of protection.

The technological transfer of knowledge will be a direct added value for the research project partner (the beneficiary company), ensuring its efficient market position - by reducing losses and accidents - which is an implicit goal, given the vast content of knowledge that INCDPM may transmit and use - in research requested by other beneficiaries or joint research with potential beneficiaries [4].

Their economic competitiveness is strengthened by the technology transfer of research results to project industrial partners and new/potential beneficiaries, which is a market advantage.

The final goal of the research project is to conduct research for the development of an intelligent decision support system for the design and development of products to ensure protection against electric shock in electrical installations, in the sense of materializing specific safety elements to minimize risks that may arise in disasters.

The research presented in this paper aims to identify the risk elements that can lead to disasters as a result of the design and manufacture of work equipment that does not meet specific safety and health requirements, as well as the legal provisions of European Directives and technical regulations, as well as national security in the field of occupational safety and health, as well as the measures to be taken to avoid these dangers.

References

[1] Bejinariu C, Darabont D C, Baciu E R, Ioniță I, Bernevig Sava M and Baciu C 2017 *Environmental Engineering and Management Journal* **16** 1395–1400
[2] Darabont D C, Bejinariu C, Ionita I, Bernevig Sava M A, Baciu C and Baciu E R 2018 *Environmental Engineering and Management Journal* **17** 2711–2717
[3] Haiducu M, Ștepa R A, Chiurtu R E and Schmitz-Felten E 2019 *Environmental Engineering and Management Journal* **18** 825–830
[4] INCDPM, 2020-2023 Operational study of the specific problems in the field of car power supply, identifying the main problems, identifying risks and consequences that may result from an unforeseen event that turns into an accident Risk structure - causes - consequences draft prevention measures. Specific ontology
[5] SR EN 61243-3: 2015 Live working - Voltage detectors - Part 3: Two-pole low-voltage type
[6] INCDPM, 2013 Guide for the application of GD No.1146 / 2006 regarding the minimum safety and health requirements for the use of work equipment by workers (in Romanian), National Institute of Research and Development for Occupational Safety Alexandru Darabont, Bucharest, Romania, Online at: http://www.inpm.ro/files/publicatii/2013-05.01-ghid-t.pdf
[7] Antonov A, Buica G, Beiu C, Darabont D, Risteiu R, Dobra and Ayaz Ahmad M 2019 *East African Scholars Journal of Engineering and Computer Sciences* pp 103–109
[8] INCDPM, 2018 Establishing the requirements and methods necessary for technical and safety verifications of protective equipment for work in live electrical installations - premise for ensuring the competitiveness of the management of companies in the energy sector, research study
[9] Buică G 2010 *Contributions of diagnostic methods of safty and health at work in electrical installations, PhD Thesis* (Petroşani: Universitas) p 70–149
[10] Dašić P, Hutanu C, Jevremović V, Dobra R, Risteiu M and Ileana I 2017 *IOP Conference Series: Materials Science and Engineering* **209** 012068
[11] Buica G, Antonov AE, Beiu C, Dobra R, Risteiu M 2018 *IOP Conf. Series: Materials Science and Engineering* **374** 012008
[12] Buica G, Antonov AE, Beiu C, Dobra R, Risteiu M 2017 *IOP Conf. Series: Materials Science and Engineering* **209** 012094
[13] SR EN 60068-2-75: 2015 Environmental testing - Part 2-75: Tests - Test Eh: Hammer tests
[14] SR EN 61851-1: 2019 Electric vehicle conductive charging system - Part 1: General requirements
[15] SR EN 62262: 2004 Degrees of protection provided by enclosures for electrical equipment against external mechanical impacts (IK code)
[16] Pasculescu D, Lupu L, Pasculescu V, Inisconi I and Suvar M 2012 *Environmental Engineering and Management Journal* **11**(7) 1267–1271
[17] Dobra R, Pasculescu D, Marc G, Risteiu M and Antonov A 2017 *IOP Conf. Series: Materials Science and Engineering* **209**
[18] Leba M, Ionica A, Dovleac R and Dobra R 2018 *Sustainability* **10**(2) 332
[19] Handra AD, Pasculescu D and Popescu FG 2020 *Annals of 'Constantin Brancusi' "University of Targu Jiu", Engineering Series* **1** 47–53
[20] Vizureanu P and Agop M 2007 *Materials Transactions* **48(11)** 3021-3023. DOI: 10.2320/matertrans.MRP2007118.
[21] Nejneru C, Cristina Perju M, Doru Burduhos Nergis D, Victor Sandu A and Bejinariu C 2019 *Revista de Chimie* **70** 3597–3602

[22] Nergis, DDB, Vizureanu P, Sandu AV, Nergis, DPB and Bejinariu C 2022 *Materials* **15(1)** 202. DOI: 10.3390/ma15010202.

[23] Tadesse MG, Nagy L, Nierstrasz V, Loghin C, Chen Y and Wang L 2018 *Materials* **11(12)** 2466. doi: 10.3390/ma11122466.

IOP Conf. Series: Materials Science and Engineering 1251 (2022) 012007 doi:10.1088/1757-899X/1251/1/012007

Investigation of microstructural development in mortars with recycled paper ash used as cement replacement, using NMR relaxometry

M L Tintisan[1], A C Siomin[1], M Moldovean[1], A M Zaharie[1], R Istoan[1], E Jumate[1], D L Manea[1] and R Fechete[2]

[1]Department of Civil Engineering and Management, Faculty of Civil Engineering, Technical Univercity of Cluj Napoca, 15, Constantin Daicoviciu Street, RO-400020 Cluj-Napoca, Romania
[2]Department of Physics and Chemistry, Faculty of Materials and Environmental Engineering, Technical Univercity of Cluj Napoca,25 George Barițiu Street, RO-400027, Cluj-Napoca, Romania

Email: coniac.loredana@gmail.com

Abstract. For the investigation of porous media with applications in the field of civil engineering on concrete and mortar, NMR technology is used, with emphasis on the sensitivity of 1H hydrogen protons to pore filling fluids. NMR technology was used to highlight the changes that replacing cement with recycled paper ash powder in mortar recipes brings at a molecular level. The samples are analyzed during 56 days, at predetermined deadlines depending on the evolution of cement compounds hydration. Ability of recycled paper ash to replace cement in its role as binder is analyzed and the results obtained from mechanical tests are correlated with the structure and volume of pores in the control samples. Mixture compositions and C–S–H characteristics are calculated based on NMR signal intensities and relaxation analysis.

1. Introduction

Sustainable or green buildings, as they are called in the literature, have gained more and more notoriety. Both beneficiaries and engineers understood the need to identify new ways to reduce their carbon footprint, energy consumption and costs involved in creating a living or working spaces. The aim of green construction is to introduce sustainable and recycled building materials, reduce waste and use renewable energy resources. According to Prof. Dr. Charles J. Kibert, the field of sustainable construction is defined by: responsible management of a healthy construction environment, based on ecological principles and resource efficiency [1]. According to the same researcher, in addition to the main traditional concerns in construction (performance, quality, cost) the main criteria that concern us in sustainable construction are resource depletion, environment degradation and environmental health.

The goal of this research is to identify a feasible method to replace a percentage of the cement used in the production of mortars with a waste generated by humanity constantly and in large quantities. The idea of ash was not new, ash from various materials has been studied over the years in laboratories, but the idea of ash from paper and other cellulosic waste was a relatively new idea, studied by other researchers, but with immense potential and with undiscovered mysteries [2-4].

2. Recycled paper ash and ash flakes mortar recipes

In this study are analyzed 3 recipes chosen based on a larger study, presented in other papers [5,6]. The standard recipe is a CS4 plaster mortar and M5C class masonry mortar, with the composition described in Table 1. RPA (recycled paper ash - Figure 1a) is used as a cement substitute, and AF (ash flakes - Figure 1b) are used as an additive, calculated in percentages by mass of cement quantity. The description of the short name of studied recipes is described in Table 2.

Table 1. Basic mortar recipe CS4 and M5C(0-2).

	Cement [g]	Sand [g]		Water [ml]
CS IV	2695	0-0.25mm	350	1446
		0.25-0.5mm	1050	
		0.5-1mm	1710	
		1-2mm	1050	
		2-4mm	2800	
M5C(0-2)	1875	0.1-1mm	4500	1310
		1-2mm	1125	

Table 2. Short name meaning and composition.

Recipe name	Recipe composition
CS IV	Standard CS4 plaster mortar
CS IV+15C	CS IV with 15% of cement quantity replaced by RPA
CS IV+15C+5F	CS IV with 15% of cement quantity replaced by RPA and AF used as additive, 5 % by mass of cement quantity is added
M5C(0.1-2)	Standard M5C class masonry mortar with 0.1-2 mm size aggregates
M5C+15C	M5C(0.1-2) with 15% of cement quantity replaced by RPA

The ash used during the experimental program is obtained by burning mixed cellulosic waste, resulting in a heterogeneous material. Because the ash used in all tests is a pre-homogenized batch during all tests, it is considered a homogeneous material with the same properties. Following the sieving, two types of material were divided, based on granular size. Granules with a size of less than 0.068 mm (cement particle size) called RPA (Figure 1a) - used as a cement substitute, and granules with a size between 0.068 mm and 4 mm called AF (Figure 1b) - used as an additive.

Figure 1. a) RPA – recycled paper ash, fine particles; b) AF – ash flakes.

3. Tests performed on mortar recipes samples

To identify the physical and mechanical properties of hybrid mortars, numerous tests were performed for each recipe. In order to establish the effect that ash has at the microscopic level in the structure of the studied mortars, modern techniques involving NMR technology were used.

3.1. Results of NMR relaxation analysis

Nuclear magnetic resonance in particular proton NMR relaxometry is an essential method for characterizing the state of water in porous materials [7-9]. The results are obtained in the form of distribution curves of the probability of finding a certain type of relaxation in a certain range. A relaxation time with low values can be associated with hydration water, while increasing values of the relaxation time t_2 capture water from small, large and medium pores [10,11]. So, usually 3 or 4 peaks can be distinguished, whose position can be associated with the three types of pores. In order to better understand the graphs with the recorded results to be presented, it can be seen in Figure 2 the schematic microscopic structure of the cement granules in the mortar mixture.

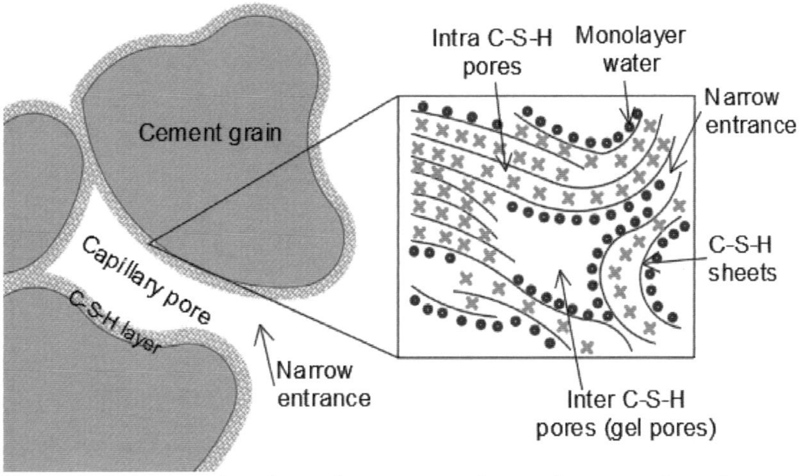

Figure 2. Microscopic schematic structure of pores in cement based mortars.

This scheme represents the cement paste, consisting of cement granules and water. C-S-H layer or calcium silicate hydrate layer, is present on the surface of unhydrated cement granules. This structure is a lamellar structure, which contains two categories of pores: inter C-S-H pores, called medium pores, intra C-S-H pores, also called small pores [12]. The intra C-S-H pores are always filled with water, because if the water were removed, the structure would lose its rigidity. Water from medium pores can be removed by drying, leaving behind a porous structure that maintains its shape and rigidity. Large pores are formed by a system of larger capillary pores. Using the CPMG method to measure the t_2 relaxation time, a graph is generated, the significance of which is explained in Figure 3. Thus, water from small pores or intra C-S-H is found in the form of the first peak. This peak has a high aptitude, and is necessarily present for each measurement, because it represents the hydration water, which is always present in the mortar structure, regardless of the state of maturation.

Thus, Figure 4 shows the distribution curves of the relaxation time t_2 measured for CSIV + 15C by the CPMG method for which the echo occurrence time was 70 microseconds. The mortars were measured at 3, 7 and 28 days after preparation. Relaxation time distributions show 3 or 4 distinct peaks. At 3 days after preparation, there is a wide distribution, centered at about 300 microseconds, and which can be associated with hydrating water, and 2 small amplitude peaks, representing mobile water, in positions that can be associated with small and medium pores. At 7 days after pouring there is a large peak, specific for hydration water, as well as 2 picks with smaller amplitudes, almost overlapping, with a location similar to the previous case, these exemplifying the water from small and medium pores. Instead, at 28 days, 4 clear peaks are observed. The largest amount of water is in the form of hydration water, it is observed by the appearance of a pick at approx. 200 microseconds. Equal amounts of water are observed by the appearance of 2 drops at a time t_2 of 3 milliseconds (small pores) and 20 milliseconds (medium pores). A small amount of water in large pores is identified by the appearance of a small amplitude peak at about 100 milliseconds.

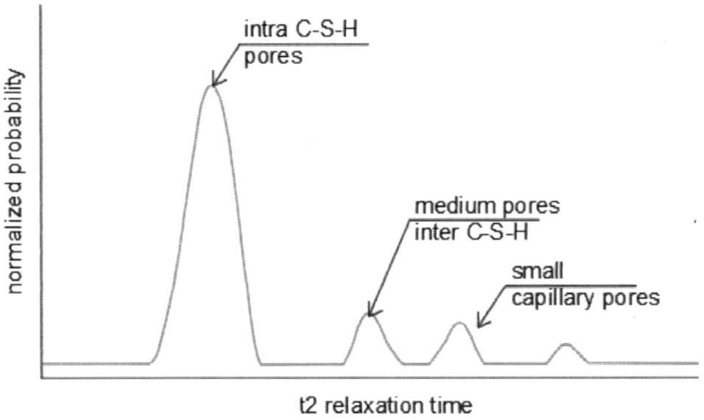

Figure 3. T_2 relaxation time distribution graph explained.

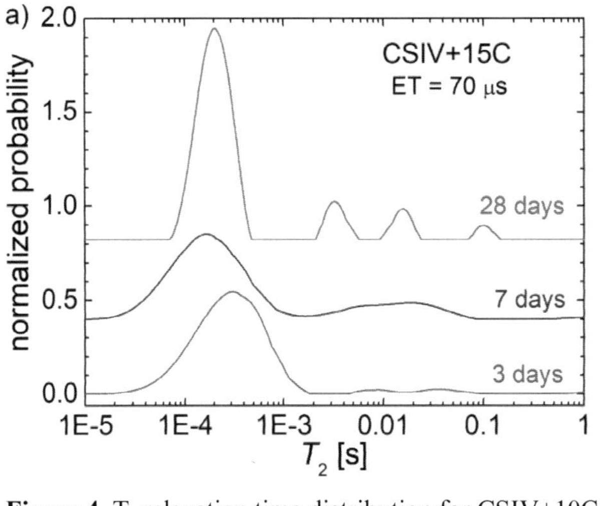

Figure 4. T_2 relaxation time distribution for CSIV+10C at 3,7 and 28 days.

Figure 5 shows the relaxation time distributions t_2 measured for CSIV + 15C + 5F. The appearance of 3-4 drops is observed. The behavior is relatively similar to CSIV + 15C. Thus, at 3 days, the largest amount of water is found as hydration water, and a quantity of water from which small and medium pores will eventually develop. At 3 and 7 days they are not well differentiated. Also in this case, at 28 days, the largest amount of water is found as hydration water, but there is also observed water from small, medium and large pores. Again, the largest amount of water is found in large pores.

Figure 6 shows the same distributions of the relaxation time t_2, but for the M5C + 15C mortar recipe. Again, a distribution of t_2 is observed 3 days after preparation, which indicates that the largest amount of water is found to be bound water. Instead, at 7 days, a significant amount of water is already found in medium-sized pores. This water is about half the amount of hydrating water. Most likely, from these average pores at 7 days, large pores will form, as seen at 28 days, which will contain the smallest amount of water, a normal effect for mortars that are considered dry at 28 days.

Significant amounts of water are in the medium sized pore, as seen for the peak located at about 15 milliseconds, and in small sized pores - the peak located at 3 milliseconds.

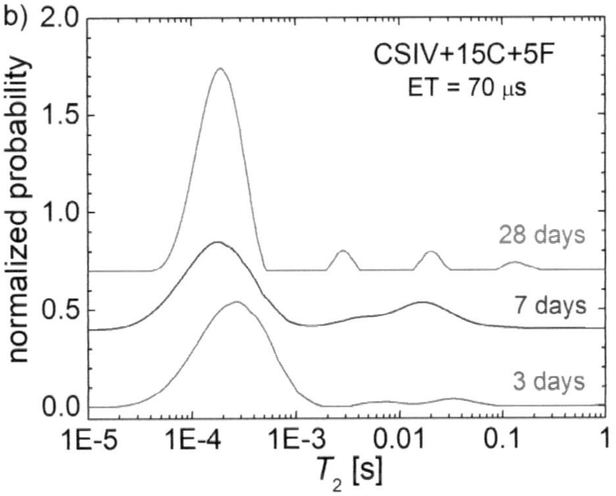

Figure 5. T_2 relaxation time distribution for CSIV+10C+5F at 3,7 and 28 days.

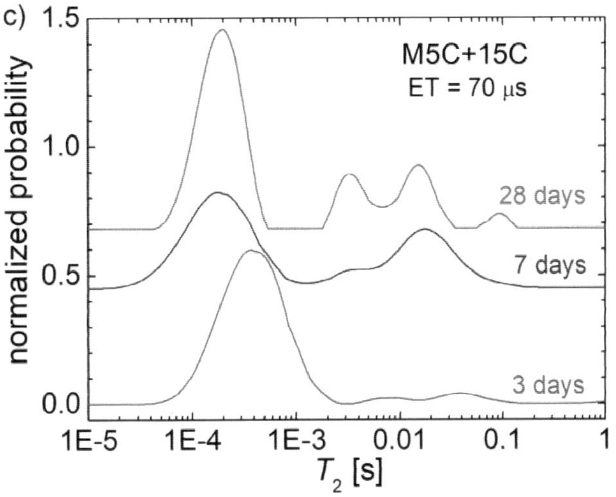

Figure 6. T_2 relaxation time distribution for M5C+15C at 3,7 and 28 days.

In Figures 7, 8 and 9, the distribution curves for the control samples are compared, at 3.7 and 28 days, in order to identify specific features of each studied recipe.

In Figure 7, the control samples are presented 3 days after casting. It is observed that the distributions of t_2 are approximately similar. Thus, the largest amount of water for all these mortars is found as hydration water, at the moment, which will probably eventually be bound water, which will lead to the stiffening of the material. According to previous studies this parameter can be associated with mechanical resistance to compression and tension. It can also be seen that in the case of CSIV + 15C + 5F mortar, the distributions extend a little more towards lower t_2 values, indicating a fraction of this water in more rigid areas but it is not yet known if this is reflected in the rigidity of the whole matter. Small amounts of water are observed in the areas where small and medium pores will develop. Basically, 3 days after preparation, these pores begin to be highlighted.

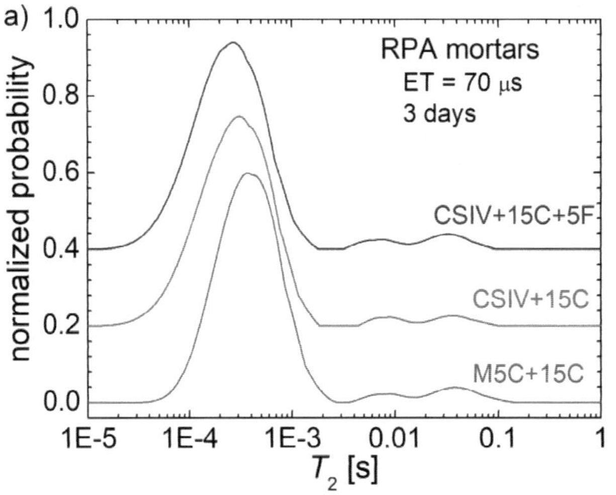

Figure 7. T_2 relaxation time distribution for CSIV+15C+5F, CSIV+15C and M5C+15C at 3 days of age.

Figure 8 shows the distributions 7 days after preparation. In this case, for the mixtures M5c + 15c and CSIV + 15C + 5F, a larger amount of water is already observed in the middle pores which are starting to be better formed. The same is not true for CSIV + 15C mortar. Because a significant amount of water is found in these medium pores, hydration water is present in a directly proportional smaller amount.

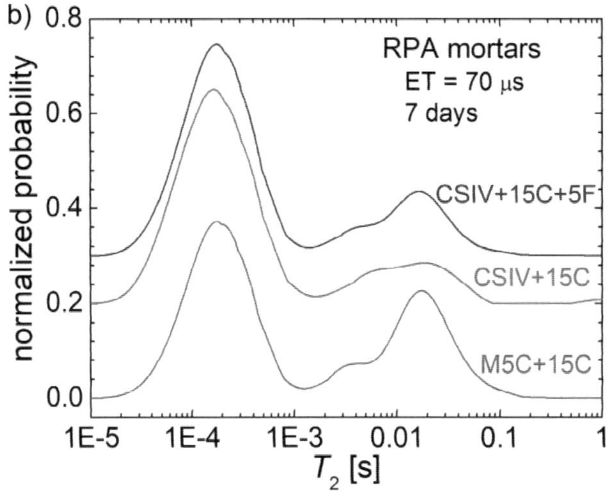

Figure 8. T_2 relaxation time distribution for CSIV+15C+5F, CSIV+15C and M5C+15C at 7 days of age.

Figure 9 shows the distributions of t_2, for the control samples, 28 days after preparation. In this case, all the peaks are well outlined, which indicates that the hydration water can be called bound water, which is closely related to its rigidity. Small, medium and large pores are already fully formed, especially for CSIV + 15C and CSIV + 15C + 5F. In the case of M5C + 15C, there is a slight overlap between the distributions of t_2, corresponding to the small and medium pores, which indicates a slight increase in inhomogeneities at this level. It is also observed that the relative amount of water found in the medium pores is higher than in the case of the other 2 mortar recipes.

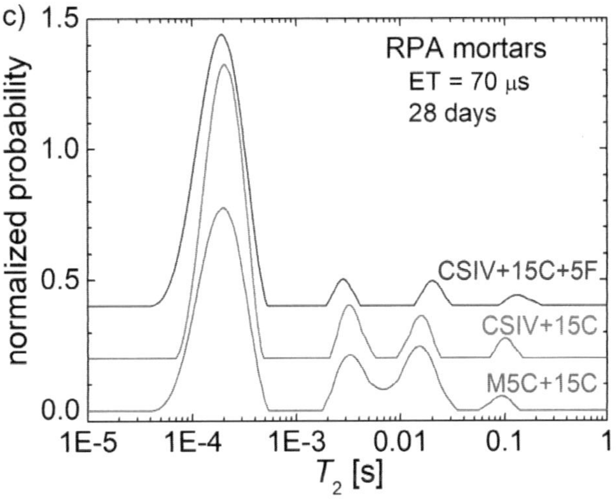

Figure 9. T_2 relaxation time distribution for CSIV+15C+5F, CSIV+15C and M5C+15C at 28 days of age.

3.2. Results of mechanical tests

In order to outline a conclusion for the study of mortars with RPA and AF, in this chapter will be presented the results of mechanical strength. For each of the studied recipes, tests were performed according to the regulations in force, at 3.7 and 28 days after pouring.

Thus, in Figure 10 is presented the variation of the bending strength, determined according to SR EN 1015: 2020-11 at 3, 7 and 28 days, for each sample. The evolution of the studied mortar recipes is analyzed compared to the standard CS4 recipe. There is a constant linear evolution of all the recipes studied during the 28 days of testing.

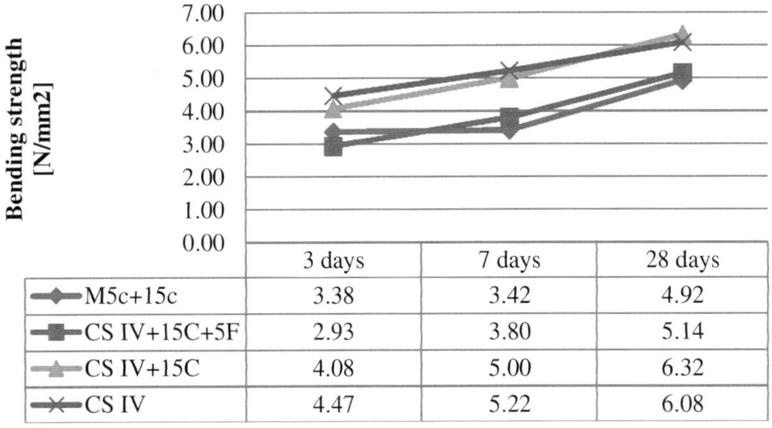

	3 days	7 days	28 days
M5c+15c	3.38	3.42	4.92
CS IV+15C+5F	2.93	3.80	5.14
CS IV+15C	4.08	5.00	6.32
CS IV	4.47	5.22	6.08

Figure 10. Bending strength evolution at 3, 7 and 28 days.

At 28 days, according to the graph in Figure 11, the CS4 + 15F mortar exceeds by 4% the tensile strength recorded by the standard recipe, a value graphically expressed by the red horizontal line. In contrast, CS4 + 15C + 5F has a value 16% lower than the standard recipe, and M5C + 15C a decrease of 20%.

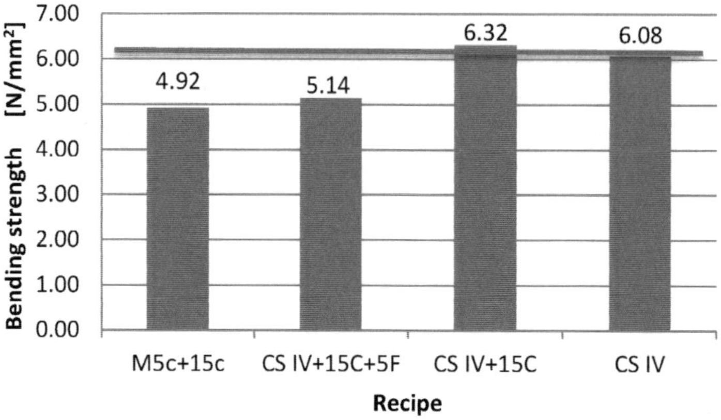

Figure 11. Bending strength results at 28 days.

The behaviour of the mortars is a linear one, according to the graph in Figure 12, the evolution having a constant trend of increasing the compressive strength during the 28 days of study. At 3 days after pouring, all mixes have lower resistance than the standard recipe, as well as at 7 days. The differences appear at 28 days where the evolution of the CS4 + 15C recipe is highlighted.

The behaviour at 28 days, represented graphically in Figure 13, is similar to that recorded in the case of bending strengths, so the most effective mix is that in which 15% of the amount of cement is replaced with RPA. The compressive strength for the CS4 + 15C mix is approximately equal to the standard recipe. The mixture with 15% of the amount of cement replaced by RPA, and 5% addition of flakes registers a decrease of 30% of the compressive strength compared to the standard recipe CS4, and M5C + 15C a decrease of 41 percentages.

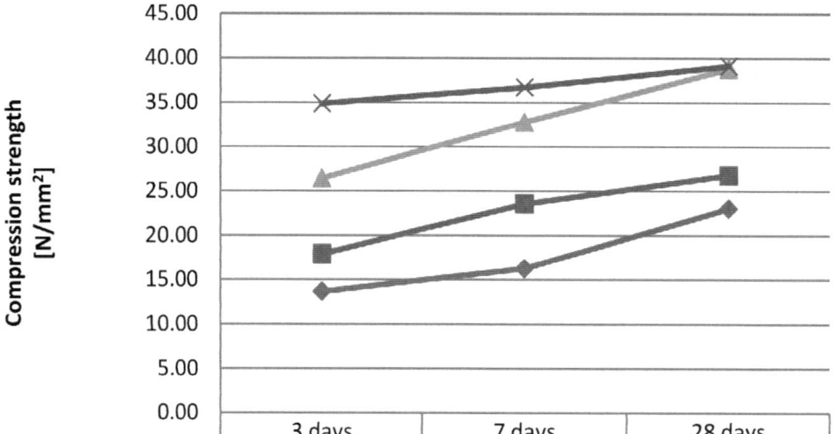

	3 days	7 days	28 days
M5c+15c	13.64	16.25	23.03
CS IV+15C+5F	17.88	23.55	26.78
CS IV+15C	26.41	32.76	38.78
CS IV	34.85	36.69	39.11

Figure 12. Compression strength evolution at 3, 7 and 28 days.

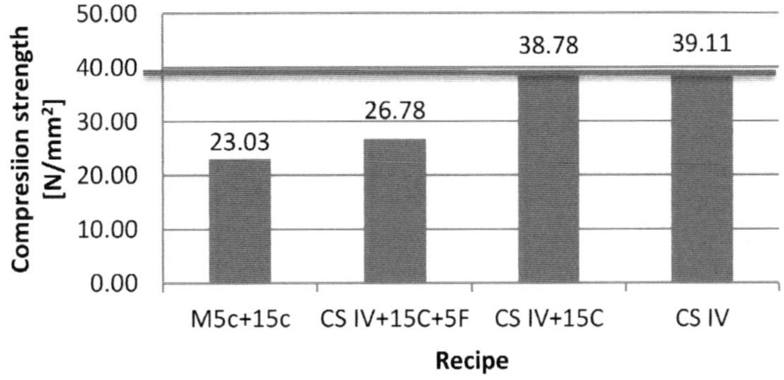

Figure 13. Compression strength results at 28 days.

4. Conclusions and discussions

The results obtained from the classical tests, such as mechanical strengths, were supported by the structural analysis of the material by the MRI relaxation technique. Thus, from the sample of studied recipes, the best results are obtained by the CS4 + 15C mixture, a cement-based plaster mortar, in which 15% of the amount of cement was replaced with RPA. In terms of both bending and compressive strength, the behaviour is superior to the CS4 standard mortar. Basically, this study demonstrates the feasibility of replacing cement with recycled paper ash, in a percentage of 15%. The addition of 5% flakes to the above mixture is not appropriate. Notable changes occur in the behaviour of the material in all 28 days of testing, the final results being clearly inferior to the standard mortar. Attempting to reduce the size of the aggregates was also a failure. M5C + 15C mortar has much lower results than the standard mortar, which is also confirmed by the distribution of small and medium pores at 28 days, which highlights an inhomogeneous behaviour, with increased presence of capillary pores.

Acknowledgment

This paper was financially supported by the Project "Entrepreneurial competences and excellence research in doctoral and postdoctoral programs - ANTREDOC", project co-funded by the European Social Fund financing agreement no. 56437/24.07.2019.

References

[1] Kibert J 2006 *Sustainable Construction: Green Building Design and Delivery* (Ireland: John Wiley & Sons).
[2] Sajad A 2013 *IOSR Journal of Engineering (IOSR-JEN)* **3** 2278-8719.
[3] Raut S, Ralegaonkar R and Mandavgane R 2013 *Archives of Civil and Mechanical Engineering* **13** 269-275.
[4] Wong H, Barakat R, Alhilali A, Saleh M and Cheeseman C R 2015 *Cement and Concrete Research* **70.**
[5] Tintisan M L, Zaharie A, Manea D L, and Siomin A C 2021 *Paper Ash Used as Substitute of cement,in Cement Mixtures,* The 15th International Conference Interdisciplinarity in Engineering. Inter-Eng, Lecture Notes in Networks and Systems (London: Springer, Cham) **386.**
[6] Tintisan M L, Siomin A C, Zaharie A, and Manea D L 2022 IOP Conf. Ser.: Mater. Sci. Eng. **1242** 012037.
[7] Fechete R, Demco A E, Moldovan D C, Chelcea R and Culea E 2010 *Rezonanta magnetica nucleara- Metode clasice si modern* (Cluj Napoca: Risoprint).
[8] Yahya Z, Abdullah MMAB, Ramli NM, Burduhos-Nergis DD and Abd Razak R 2018 *IOP Conf. Ser.: Mater. Sci. Eng.* **374** 012068. DOI: 10.1088/1757-899X/374/1/012068.

[9] Baltatu MS, Vizureanu P, Cimpoesu R, Abdullah MMA and Sandu AV 2016 *Revista de chimie* **67(10)** 2100-2102.
[10] Axinte M, Nejneru C, Perju M C, Cimpoeşu N and Hopulele I 2011 Tehnomus New Technologies and Products in Machine Manufacturing Technologies **18** 181-184.
[11] Javadi Toghchi M, Campagne C, Cayla A, Bruniaux P, Loghin C, Cristian I, Burgnies L and Chen Y 2019 *Synthetic Metals* **251** 75-84. doi: 10.1016/j.synthmet.2019.03.026.
[12] Burduhos-Nergis D-P, Sandu A-V, Burduhos-Nergis D-D, Darabont D-C, Comaneci R-I and Bejinariu C 2019 *MATEC Web of Conferences* **290** 12004.

Plant available potassium and phosphorus in arable soil: a comparative study on testing methods

A Surleva[1,*], and L Angelova[1]

[1]University of Chemical Technology and Metallurgy, 8, St. Kliment Ohridski Blvd., Sofia 1756, Bulgaria

Email: surleva@uctm.edu

Abstract. Determination of soil fertility status is based on determination of plant available nutrients which occur in several forms. This strongly challenges the choice of extracting solution and detection technique. Deriving a mathematical model for data recalculation would allow relating the results of different methods without loss of agronomical validation data. This comparative study was aimed at assessing the relationship between two methods for determination of plant available potassium and phosphorus in arable soil. Soil K and P were extracted followed a diluted double acid (Mehlich 1) and an acetate/lactate methods (ALM). Potassium in the obtained extracts was determined by flame AES and phosphorus - by molybdenum blue spectrophotometry. To compare the results obtained by the both extraction methods, the uncertainty of measurement of each method was assessed following the single laboratory validation and quality control approach. The expended uncertainty was found to be 2 mg/100g as P_2O_5 and 7 mg/100g as K_2O (at k=2). The results from the comparative analysis of 93 arable soil samples showed a correlation between Mehlich-1 and ALM extracted P and K within uncertainty limits (at k=3). Based on the recalculated values of extractable K and P, the fertility status of 60 % of soil samples were correctly classified, 30 % with acceptable error and 7 % could not be classified. The study showed that Mehlich-1 and ALM methods provided comparable results for studied range of soil samples. However, determination of soil fertility status based on recalculated results should be made with caution as it could be highly influenced by soil type.

1. Introduction

Estimation of arable soil fertility status includes determination of main nutrients: phosphorus, potassium and mineral nitrogen, as well as active acidity, organic carbon and some other characteristics. Potassium and phosphorus exist under different forms in soils and only partially are accessible to plants. The forms denoted as "plant available" nutrients present particular interest for agronomists in their efforts to develop fertilisation programs corresponding to sustainable agriculture. In order to estimate plant available P and K a line of extracting mixtures have been developed and it is well recognised that any universal extractant exists. The extracting mixtures were proved to extract different parts of soil nutrients which make the comparison of the data from different methods a complex task [1-5]. Additional difficulty arises from the used detection techniques which respond to different forms of nutrients in soil extract [6,7]. It should be pointed out that agronomical validation and indices for estimation of soil fertility status are developed based on soil testing results with specific extracting mixture and detection finishing. Possibility to transfer data from one method to another without losing agronomically validated data is in the focus of many studies [3-5,8,9]. Additionally, multi-element methods are preferable choice for soil testing laboratories and their transfer functions are of special interest [1-5,10]. Breure et al. [10] proposed transfer functions between P-Olsen and P-Mehlich 3 as well as between K-ammonium acetate and K-Mehlich 3 for soils from tropical and temperate areas. They reported that the relations between K and P extracted by studied extractants differed for both types of soils. The transfer functions included different parameters and should be evaluated separately. Reijneveld et al. [8] compared near infrared spectroscopy and multi-nutrient 0.01 M $CaCl_2$ extraction with conventional techniques applied in Netherlands and reported good correlation between obtained results. Zehetner at al. [4] reported correlation study of soil phosphorus obtained by 14 extraction methods applied to agricultural soils from Central Europe. They found that neutral salt solutions best predict plant-available P with some limitation for calcareous soils. A line of studies aimed at comparison between extraction methods for plant available K in arable soils from different regions with conversation between soil test results have been recently reported [11-14]. Many researchers studied soil multi-element extracting solutions to determine some nutrients availability for crops in order to decrease analysis costs and labor [5,13,15,16]. Zebec et al [5] demonstrated remarkable correlation between the results obtained by different extraction approaches in soil with different characteristics. Rogers et al. [12] proposed specific correlation equations that could be used for comparison among tests for extractable K determination in soils from US, Idaho region. A modified multi-nutrient acetate-lactate method for determination of plant available K and P was especially designed for Bulgarian arable soils with wide range of characteristics [17,18]. The instrumental detection for K was flame atomic spectrometry and UV-Vis spectrophotometry for P. The method was specifically calibrated for Bulgarian arable soils for a line of agricultural plants [18-20]. Analytical behaviour and uncertainty estimation of the modified method were previously reported by our team [21,22]. The results showed that the soil sample inhomogeneity and the repeatability of extraction process were the main factors which contribute to the uncertainty of measurement in soil analysis. Comparison between some of the widely applied methods for K determination in Bulgarian arable soils revealed that results from ammonium acetate and acetate-lactate extractions were highly correlated, acetate-lactate and Mehlich-1 extraction resulted in satisfactory correlation with Mehlich-1 extracted less potassium [21]. Although comparative studies on different multi-element methods for plant available K and P have been intensively reported, the uncertainty of measurement inherent to the testing methods is rarely included [7,21,23]. As Sikora et al. [23] warned in transferring data from one soil test to another, two associated errors should be evaluated. The first contributor to uncertainty arises from obtaining the original result from soil analysis, and the second contributor - from transfer equations. Thus, estimation of uncertainty of methods applied for soil testing should benefit the further research on transfer functions.

The aim of this study was to explore two multi-nutrient soil testing methods for estimation of plant available potassium and phosphorus in arable soils based on a diluted double acid (Mehlich-1) and an acetate-lactate extraction, followed by spectrophotometry for phosphorus and flame atomic emission spectrometry for potassium determination is soil extracts, as well as to reveal the correlation between the results obtained by both methods.

2. Experimental

Preparation of Murphy-Riley colour reagent: A premixed Murphy-Riley solution was prepared as follows: 12 g of $(NH_4)_6Mo_7O_{24}.4H_2O$ were dissolved in 500 mL of d. H_2O and acidified by 50 mL 10 M HCl; 0.2928 g of $C_8H_4K_2O_{12}Sb_2.3H_2O$ were dissolved in 500 mL water; both solutions were added to 1 L 2.5 M H_2SO_4 and diluted up to 2 L. Thus prepared solution was stored in dark at room temperature (20 °C). The solution was stable at least two weeks. Molybdenum blue colour reagent was prepared daily by dissolving 1.056 g ascorbic acid in 200 mL of the above solution. All reagents were of p.a. grade, Sigma-Aldrich, deionized water was used.

Preparation of a modified acetate-lactate extracting solution [17,21]: A modified ammonium acetate – calcium lactate extracting solution at pH=4.2 was prepared by dissolving 154.2 g of $C_6H_{10}CaO_6.5H_2O$ at gentle heating; after cooling, 50 mL of 10 M HCl were added and the solution was made up to 1 L with d. H_2O; 77,1 g CH_3COONH_4 were dissolved and mixed with 60 mL of glacial acetic acid; the solution was made up to 1 L; both solutions were mixed and diluted to 5 L with deionized water. All reagents were of p.a. grade, produced by Sigma-Aldrich. The pH of the solution was measured with pH meter equipped with a combined glass electrode. pH was corrected by aqueous ammonia or acetic acid additions.

Plant available phosphorous extraction [17,21]: Soil samples were air dried, grounded and 2 mm sieved. Plant available phosphorous and potassium were extracted from arable soil applying two procedures: a) a modified acetate-lactate extractant. The extraction was carried out for 1 hour at soil-to-liquid ratio 1:25 (2 g accurately weighted dry soil was mixed with 50 mL extracting solution). b) diluted double acid extraction (Mehlich-1). The extracting solution contained 0.05 M HCl and 0.0125 M H_2SO_4. The soil-to-solution ratio was 1:5 (10 g of accurately weighted soil samples were mixed with 50 mL of extracting solution for 5 min). The samples were filtered through paper filter and the filtrates were subjected to phosphorous determination by spectrophotometry and potassium determination by flame AES. Each soil sample was analysed in duplicates and the mean content of the studied nutrients was calculated.

Determination of phosphorous by spectrophotometry [17,22]: The extracted phosphorus was quantified by the molybdenum blue spectrophotometric method. A reference material 50 mg/L PO_4^{3-} (Hach-Lange, Lot A0080, traceable to NIST) was used for preparation of calibration standards by appropriate dilution in the concentration range: 0.1-1.0 mg/L as P_2O_5. Two mL of extracting mixture was added to each standard solution to mimic soil extracting medium. The blank was prepared by adding 2 mL extracting reagent, 4 mL MB colour reagent and diluted up to 25 mL with d. H_2O. The absorbance of each solution at 880 nm was measured in triplicate against the reagent blank after one hour for color development using Hach Lange DR 3900 Spectrophotometer. The mean absorbance value was used for standard curve construction and the calibration equation was obtained applying linear regression by the least squares method to the obtained data. For determination of phosphorous in soil extracts, 2 mL of the extract were mixed with 4 mL of MB colour reagent and volume was made up to 25 mL in a measuring flask. After 1 h, the absorbance was measured at 880 nm against a reagent blank.

Potassium determination by flame atomic emission spectrometry: The extracted K was determined by Jenway PFP7 Flame photometer equipped with interference type colour filter and PIN diode detector [23-25]. The intensity of the emitted light was measured at 766 nm. Low temperature air/ propane-butane gas mixture was used at the following conditions: airflow of 6 L/min at 12 psi, sample uptake 4.5 mL/min, daily adjusting of propane-butane gas flow to obtain a maximum signal for nominal K standard. The concentration of the nominal standard was chosen depending on the extracting procedure and expected sample concentrations [26-27]. The instrument zero was adjusted using the studied extracting solutions. Individual calibration curve was constructed for each extracting solution. The instrument performance was monitored by the readout for the nominal standard solution. A certified reference material (CRM) containing 1000 mg/L (TraceCert, Lot BCBV7454- SigmaAldrich) was used for preparation of calibration standards by appropriate dilution with extracting solutions at the following concentration range: 5 - 20 mg/L K in acetate-lactate extracting mixture and 5 - 40 mg/L in Mehlich-1. The studied extraction mixtures were used as instrument blanks. The analytical function was obtained applying polynomial regression by the least squares method.

3. Results and Discussion

3.1. Uncertainty estimation and quality control

The uncertainty was estimated following the single laboratory validation and quality control approach [28,29]. Soil test samples were used as internal quality control samples and were analysed during each bath of 12 samples. The measurands were extractable potassium and phosphorus concentrations in 100 g dry soil sample, expressed in mg K_2O/100g or mg P_2O_5/100g. Quality control samples at two concentration levels of K and P were analysed in duplicates and the mean values were used for control chart construction. The soil samples contained: mineral N 14.8 and 39.5 mg/kg, P_2O_5 4.6 and12.5

mg/100g, K_2O 20.9 and 46.8 mg/100g and organic carbon 2.96 and 4.55 %, pH 7.24 and 7.44, respectively.

The precision was estimated in within lab reproducibility conditions by standard deviation of concentration values obtained during one-year period by different analysts with different reagents, but the same instruments. The results are presented in Table 1. The control charts at two concentration levels for each of studied plant nutrients: 4.6 and 12.0 mg/100g P_2O_5, as well as 24.0 and 49.8 mg/100g K_2O, respectively, are presented on Figures 1 and 2. The warning limits were set at 1*s and the action limits at 2*s.

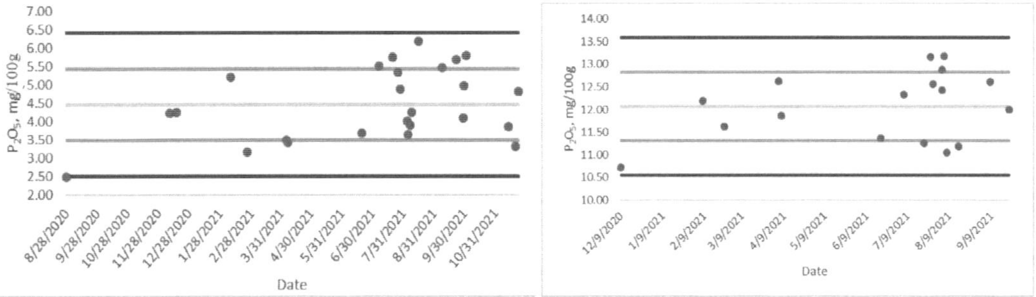

Figure 1. Control chart of arable soil sample containing (a) 4.6 and (b) 12.1 mg/100 g P_2O_5. The warning limits (green lines) were set at 1*s and the action limits (red line) at 2*s, blue line presented the mean value.

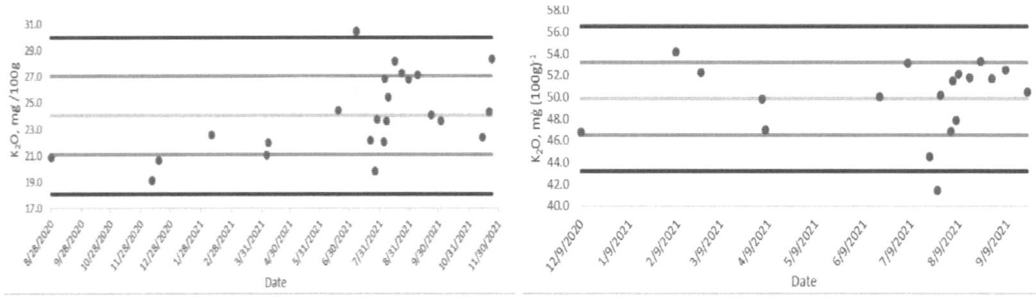

Figure 2. Control chart of arable soil sample containing (a) 24.0 and (b) 49.8 mg/100 g K_2O. The warning limits (green lines) were set at 1*s and the action limits (red lines) at 2*s, blue line presented the mean value.

The uncertainty was estimated according to the Equation (1) [29]:

$$u_c = \sqrt{u(s)^2 + u(bias)^2} \quad (1)$$

where: u_c was combined uncertainty, u(s) was the standard uncertainty calculated as standard deviation from the control chart (n=25) and u(bias) was the standard uncertainty of the bias calculated by standard deviation of bias (n=3). The bias was determined by standard addition of analyte RM to the soil extract.

The results are presented in Table 1.

Using control charts for uncertainty calculation allowed estimating the influence of as much as possible random factors on the analytical result thus reflecting within lab reproducibility at real working practice rather than specially designed plan.

The advantage of the quality control approach is the transformation of a systematic factor (calibration curve) into random one, when applying everyday calibration procedure [30-33]. However, this approach did not reveal the contribution of individual factors to the measurements uncertainty. The results showed that expanded uncertainty was 2.2 mg P_2O_5/100 g and 7.0 mg K_2O/100 g at coverage factor 2.

Table 1. Uncertainty calculation by a single laboratory and quality control approach (each individual result is a mean of duplicate samples for a given day).

soil sample	concentration	within lab reproducibility		bias	uncertainty	
		s	n	s	combined u_c	expanded U (k=2)
		plant available P as P_2O_5/100 g dry soil				
1	4.5	0.81	19	0.43	0.9	1.8
2	12.1	1.01	24		1.1	2.2
		plant available K as K_2O/100 g dry soil				
1	24.0	2.97	19	1.1	3.2	6.4
2	49.9	3.32	24		3.5	7.0

Detailed study of the repeatability of P determination by spectrophotometry based on molybdenum blue color reaction (Table 2) demonstrated that the main contribution to the uncertainty of the studied method for plant available P determination was the sample inhomogeneity sieved at 2 mm.

Table 2. Repeatability data of extractable soil phosphorus determination by molybdenum blue method after extraction with acetate-lactate reagent.

sample	Absorbance/Concentration			N	mean	s	s_r, %
	repeatability of measurement of absorbance						
	A	A	A		A_{mean}		
1	0.145	0.146	0.145		0.145	0.0006	0.4
2	0.286	0.289	0.289	3	0.288	0.0017	0.6
3	0.577	0.577	0.576		0.577	0.0006	0.1
	repeatability of MB reaction and measurement of absorbance						
	A	A	A		A mean		
1	0.087	0.091	0.089		0.089	0.0020	2.2
2	0.127	0.126	0.127	3	0.127	0.0006	0.5
3	0.262	0.265	0.261		0.263	0.0021	0.8
	repeatability of extraction, MB reaction and absorbance measurement						
	C, mg P_2O_5/100 g		-		C mean		
1	12.2	11.5	-		11.9	0.50	0.92
2	6.1	6.9	-	2	6.5	0.53	2.02
3	8.2	8.3	-		8.3	0.02	0.07

The repeatability of K determination in soil samples (RSD 3 %) [21] was comparable to 4 % for P determination [22]. Hence, the extraction of both nutrients was influenced in the same manner by sample particle size, sample inhomogeneity and small variation in extraction conditions.

3.2. Comparative study of extractable P and K by Mechlih-1 and acetate-lactate method

Ninety-three real soil samples were subjected on P and K determination after extraction by Mehlich-1 and acetate-lactate methods. The characteristics of the studied soils are presented in Table 3.

Table 3. Characteristic of the studied soils.

	pH	Organic C, %	Exchangeable P_2O_5, mg / 100 g	Exchangeable K_2O, mg / 100 g	ΣNO_3^- $+NH_4^+$, mg/kg	CEC, $cmol_c$/ kg	EC, µS/cm	Clay fraction, %
max	8.66	6.00	40.7	67	101	39.0	1214	56
min	4.54	0.70	0.4	5.4	14	17.4	97.2	15
median	6.79	2.40	2.1	35	30	27.0	170.4	26
mean	6.87	2.49	4.8	35	37	28.3	241.3	27
N	93	93	45	93	77	24	24	24

The results from the correlation study between the P and K content obtained by diluted double acid (Mechlih-1) and the modified ammonium acetate - calcium lactate methods applied to arable soils are presented on Figure 3.

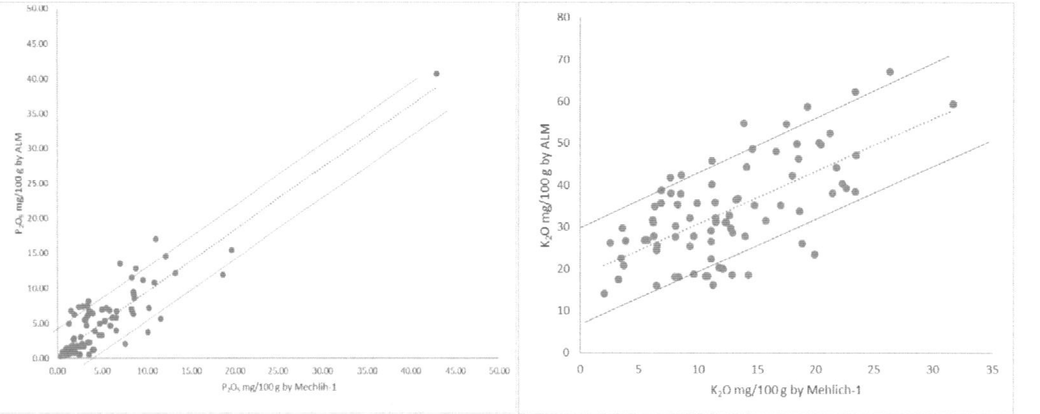

Figure 3. Comparison of extractability of diluted double acid (Mehlich-1) and a modified acetate-lactate (ALM) reagents for determination of plant available soil P and K from soil samples collected at depth of 0 to 30 cm from arable fields in Bulgaria. The dotted line represents the linear regression fit and the solid lines - uncertainty interval (U= 4.5 mg P_2O_5/100 g and U= 10 mg K_2O/100 g at k=3).

The presented results have showed that the concentrations obtained by both methods corresponded within the uncertainty limits. Soil phosphorus extracted by Mehich-1 and ALM was correlated according to the equation y=0.89*x+0.60 with R^2=0.80, n=45, 2.5 % of the results were removed. The Mehlich-1 method extracted lower quantities of P compared with ALM. Soil potassium extracted by Mehich-1 and ALM correlated by y=1.30*x+18.4 (R^2=0.40, n=93), 2 % of the results were removed. The Mehlich-1 method extracted lower quantities of K compared with ALM. The correlation equations were derived used the least squares method. By applying the correlation equations, the concentrations of P_2O_5 and K_2O were calculated from the experimental data obtained by Mehlich-1 method. Thus, recalculated values for concentrations by acetate/lactate method were obtained. Based on the obtained recalculated concentrations, the soil fertility status was estimated applying the fertility indices for ALM (Table 4). For comparison the fertility indices for Mehlich-1 obtained data are presented in the Table 4 and it could be seen that the Mehlich-1 indices levels are lower than acetate/lactate levels [18,28]. It corresponded well with the findings that Mehlich-1 extracted lower quantities of both P and K in the studied soils.

Comparing the fertility status of the studied soils based on experimental and calculated indices for plant available phosphorus, it was found that 76 % of samples the estimated fertility status coincided with experimental one, 22 % - differed by one level of fertility, 1 % differed more than one level and the fertility could not be estimated based on recalculated values. The results are presented on Figure 4.

Considering the extractable potassium in the studied soils, it was observed that 60 % of samples the fertility status based on recalculated values coincided with the fertility level based on the experimental ones, 33 % different by one level of fertility, and 7 % could not be estimated.

Table 4. Soil fertility indices based on acetate/lactate method and Mehlich-1.

	fertility indices					ref
	very low	low	medium	high	very high	
	Exchangeable P_2O_5, mg / 100 g					
ALM	<6.0	6.1-12.0	12.1-18.0	18.1-28.0	>28.0	[18]
Mehlich 1	< 1.5	1.6-3.2	3.2-5.3	5.3-10.3	>10.3	[28]
	Exchangeable K_2O, mg / 100 g					
ALM	<12	13-16	17-28	29-38	>38	[18]
Mehlich 1	< 1.8	1.8-4.2	4.2-8.4	8.4-16	>16	[28]

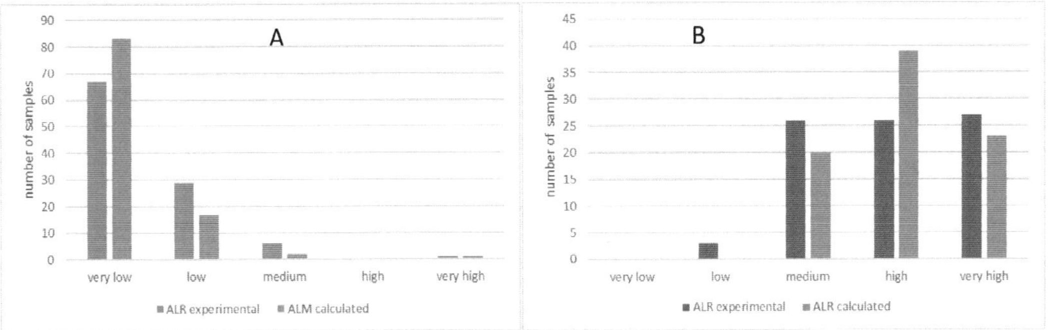

Figure 4. Comparison of soil fertility status based on experimental and recalculated concentration of P_2O_5 (A) and K_2O (B) by acetate/lactate method (ALM). Calculated values were obtained from concentrations of P_2O_5 or K_2O by Mehlich-1 method and the derived correlation equations between ALM and Mehlich-1. Blue column – experimental values by ALM. Orange columns –recalculated values Mehich-1 to ALM.

The results from recalculation of the soil testing results from Mehlich-1 to acetate/lactate method demonstrated the possibility for application of derived equations for estimation of soil fertility status based on ALM indices and data from soil analysis by Mehlich-1 method. Considering extractable phosphorus more than 75% of soil samples were successfully diagnosed. The potassium fertility status of 60 % of soil samples was appropriately estimated. However, the recalculation strategy should be applied with great caution as the uncertainty of the obtained results increased and highly depend on the soil type. The uncertainty of the derived mathematical model should be added to the to the uncertainty of measurement.

4. Conclusions

The relationship between two methods for determination of plant available potassium and phosphorus in arable soils was assessed. Soil K and P were extracted followed a diluted double acid (Mehlich 1) and an acetate/lactate methods (ALM). The uncertainty of measurement of each method was assessed following the single laboratory validation and quality control approach. The uncertainty was found to be 2 mg/100 g as P_2O_5 and 7 mg/100 g as K_2O. The results from the comparative analysis of 93 arable soil samples showed a correlation between Mehlich-1 and ALM extracted P and K within uncertainty limits set at k=3. Based on the recalculated values of extractable K and P, the fertility status of both nutrients K and P of 60 % of soil samples were correctly classified, 30 % with acceptable error and 7 % could not be classified. The study showed that Mehlich-1 and ALM methods provided comparable results for studied range of soil samples. However, determination of soil fertility status based on recalculated results should be made with caution as it could be highly influenced by soil type.

Acknowledgment
This study has been financially supported by the Science and Research Program of the university of Chemical Technology and Metallurgy, Bulgaria. The authors acknowledged the support of "Sembodja" LtD for soil sampling and valuable discussions.

References

[1] Lumbanraja J, Mulyani S, Utomo M, and Sarno 2017 *J Tropical Soils* **22** 67
[2] Medeiros M N, Oliveira F H T, Welka Preston W, et al. 2021 *Revista Ciência Agronômica* **52** e20207633
[3] Nawara S, Van Dael T, Merckx R, et al. 2017 *Eur J Soil Sci* **68** 873
[4] Zehetner F, Wuenscher R, Peticzka R, and Unterfrauner H 2018 *Plant Soil Environ* **64** 192
[5] Zebec V, Rastija D, Loncaric Z, et al. 2017 *Eurasian Soil Sci* **50** 1420
[6] Cade-Menun B J, Elkin K R, Liu C W. et al. 2018 *Geochem Trans* **19** 7
[7] Ivanov K, Zaprjanova P, Petkova M, Stefanova V, Kmetov K, Georgieva D, and Angelova V 2012 *Spectrochim Acta B* **71-72** 117
[8] Reijneveld J, van Oostrum M, Brolsma K, Fletcher D, and Oenema O 2022 *Agronomy* **12** 191
[9] Hartmann T, Wollmann I, You Y and Müller T 2019 *Agronomy* **9** 29
[10] Mirjam S, Breure M, van Eynde E, Kempen B. et al. 2021 *Geoderma* **406** 115458
[11] Madaras M, and Koubová M 2015 *Plant Soil Environ* **61** 234
[12] Rogers C, Dari B, and Schroeder K 2019 *Agrosystems, Geosciences & Environment* **2** 190067
[13] Zhang Y, Nachimuthu G, Mason S, McLaughlin M, McNeill A, and Bell M 2017 *Sci Rep* **7** 11391
[14] Li X, Zhang Y, Wang W, Khan M R, Cong R, and Lu J 2018 *Sci Rep* **8** 16381
[15] Bortolon L, and Gianello C 2010 Rev. Bras. de Ciencia do Solo **34** 125
[16] Hosseinpur A R, and Samavati M 2008 Com Soil Sci Plant Anal **39** 1559
[17] Ivanov P 1984 Soil Sci Agrochem **XIX** 88
[18] Nikolova M, Fixen P, and Popp T Best management practices for sustainable crop nutrition in Bulgaria, BMPSCN Manual on Soil and Plant Analysis and Fertilizer Recommendation. 2014, IPNI-2008-BGR-1, Sofia, Bulgaria (in Bulgarian)
[19] Hristov B, Nikova I, and Andreeva N 2017 *Bul J Soil Sci* **2** 123
[20] Zlatareva E, and Nikolov N 2015 *New know J Sci* 152
[21] Angelova L, Genova N, Stoyanova S, Surleva O, Nekov I, Ilieva D, and Surleva A 2021 *Analytics and Control* **25** 182
[22] Angelova L, Genova N, Pencheva G, Statkova Y, Violeta Yotova, and Surleva A 2022 Methods Object Chem Anal **2**
[23] Sikora F, Mylavarapu R, Hardy D, Tucker M, and Franklin R 2005 Conversion equations for soil test extractants: Mehlich-1 and Mehlich-3, http://soils.rs.uky.edu/m3vsm1.php (last accessed 10 April 2022)
[24] Axinte M, Nejneru C, Perju M C, Cimpoeşu N and Hopulele I 2011 Tehnomus New Technologies and Products in Machine Manufacturing Technologies **18** 181-184.
[25] Ciubară A, Burlea SL, Axinte M, Cimpoeşu R, Chicet DL, Manole V, Burlea G and Cimpoeşu N 2018 *IOP Conference Series: Materials Science and Engineering* **444(3)** 032013.
[26] Vizureanu P, Samoila C and Cotfas D 2009 *Environmental engineering and management journal* **8(2)** 301-306. DOI: 10.30638/eemj.2009.043.
[27] Nergis D P B, Cimpoesu N, Vizureanu P, Baciu C and Bejinariu C 2019 *Materials Today: Proceedings* **19** 969–978

[28] Nergis, DDB, Vizureanu P, Sandu AV, Nergis, DPB and Bejinariu C 2022 *Materials* **15(1)** 202. DOI: 10.3390/ma15010202.

[29] Bettencourt Da Silva R, Bulska E, Godlewska-Zylkiewicz B, Hedrich M, Majcen N, and Magnusson B. Analytical measurement: Measurement uncertainty and statistics [Joint Research Center, European Commission, 2013, pp 58-146

[30] Eurachem Guide: The Fitness for Purpose of Analytical Methods - A Laboratory Guide to Method Validation and Related Topics, Magnusson B and Örnemark U (eds.) 2014 2nd ed.

[31] Tadesse MG, Nagy L, Nierstrasz V, Loghin C, Chen Y and Wang L 2018 *Materials* **11(12)** 2466. doi: 10.3390/ma11122466. Kruve A, Rebane R, Kipper K, Oldekop M, Evard H, Herodes K, Ravio P, and Leito I 2015 *Anal. Chim. Acta* **870** 29

[32] Kruve A, Rebane R, Kipper K, Oldekop M, Evard H, Herodes K, Ravio P, and Leito I 2015 *Anal. Chim. Acta* **870** 8

[33] Heckman JR 2006 Soil Fertility Test Interpretation- Phosphorus, Potassium, Magnesium, and Calcium. Rutgers New Jersey Agricultural Experiment Station Cooperative Extension

Identifying the structure of a plaster mortar mixtures on the basis of NMR investigations

A C Siomin[1,*], M L Tintisan[1], A Zaharie[1], M Moldoveanu[1], D L Manea[1], E Jumate[1] and R Fechete[2]

[1]Department of Civil Engineering and Management, Faculty of Civil Engineering, Technical University of Cluj Napoca, 15 Constantin Daicoviciu Street, RO-400020, Cluj-Napoca, Romania
[2]Department of Physics and Chemistry, Faculty of Materials and Environmental Engineering, Technical University of Cluj Napoca, 25 George Baritiu Street, RO-400027, Cluj-Napoca, Romania

Email: cristian.siomin@googlemail.com

Abstract. Starting from the concept of sustainable development, the research program aims to establish a waste management by reducing the construction materials used in the preparation of plaster mortars. Thus, the chosen material, respectively the used paint plays an important role in replacing the main components of the plaster mortar recipe. Following the research in the laboratory of the Faculty of Construction, it was concluded that used paint can replace water or cement in the recipe in different percentages, a fact recorded in an article published in early 2022 [1]. In the future, this research can help reduce the raw materials used to make plaster mortars, an ambitious plan and a step towards the transition to the circular economy. By reducing the amount of raw materials by recycling used paint, we protect the environment. During the investigations, the 3 mortar recipes: CSIV, CSIV10C, CSIV20P are presented from the point of view of nuclear magnetic resonance NMR. According to the research program, the tracking of behavior over time was done by analyzing the samples at 3 days, 7 days and 28 days.

1. Introduction

The experimental program developed within the laboratory of the Faculty of Constructions was based on the concept of sustainable development [2,3]. Starting from this concept, plaster mortar recipes were made with the aim of using the waste resulting from the construction materials used on site, respectively the replacement of the main components of the mortar with used paint.

By pouring the plaster mortar recipes, it was established that starting from the basic CSIV recipe, the best performing recipes are the following: the CSIV 20P plaster mortar recipe in which the water was replaced with a percentage of 20% with used paint and CSIV 10C mortar recipe in which a 10% percentage of cement was replaced with used paint.

Given that the results were very promising, research continued in terms of nuclear magnetic resonance in the Department of Physics and Chemistry for the three recipes of plaster mortar: CSIV, CSIV 10C and CSIV 20P.

2. Nuclear magnetic resonance NMR investigations

Nuclear magnetic resonance NMR is one of the modern methods, most often used in the study of porous materials, respectively in this case in the study of mortars [4].

Thus, after recording an NMR signal that falls over time and can be considered multi-exponential, by a reverse Laplace analysis can be obtained the distributions of transverse relaxation times that characterize porous materials in terms of pore size, but also other characteristics of hydrogen-containing water that can bind to various components of these materials.

According to the research program, the tracking of behavior over time was done by analyzing the samples at 3 days, 7 days and 28 days.

2.1. NMR investigations on the CSIV plaster mortar recipe

In Figure 1 we have the distributions of T_2 relaxation times that show three peaks and it is observed that the largest amount of hydrogen is probably found in bound water, as seen from the appearance of a main peak, centered at about 200 microseconds to 100 microseconds. Thus, the main drop obtained for the standard CSIV mortar measured 3 days after preparation is found at about 200 microseconds, and as it hardens there is a shift of the maximum of this drop to a lower value of the relaxation time T_2, which indicates that hydrogen is found in more rigid components. This can also be a measure of a mechanical property such as compressive strength and bending strength [4-6]. There are two other peaks, quite wide, centered at about 7 milliseconds and which can be associated with water from small to medium pores and centered at about 35 milliseconds and which can be associated with water from medium-sized pores. See the curve in Figure 1 of the T_2 relaxation time at 3 days.

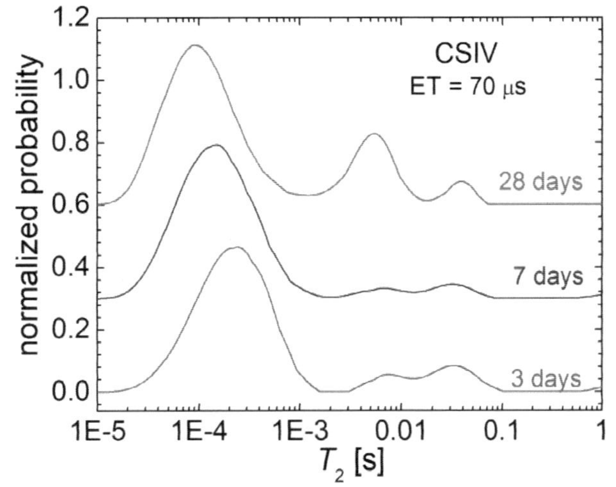

Figure 1. Normalized distribution of T_2 for the CSIV mortar recipe.

At 7 days after the preparation of the CSIV mortar, the distribution of the relaxation time on T_2 is similar to the distribution measured at 3 days except that the drops corresponding to the water from small to medium and medium pores have a smaller amplitude which indicates that probably a certain [6,7]. The amount of water has evaporated, and the fact that they are quite large indicates that the microstructure is quite inhomogeneous. At 28 days after preparation, the standard CSIV mortar has three peaks. It is observed that their relative amplitude is higher in the sense that the area under the drop corresponding to the water from the small pores to the medium and the integral area under the drop corresponding to the water corresponding to the medium pores increased. Thus, even at 28 days, the largest amount of hydrogen is found in bound water, but we also find water that is found in small to medium and medium pores.

2.2. NMR investigations on the CSIV 10C plaster mortar recipe

Below, Figure 2 shows the relaxation time distributions for CSIV 10C mortar for which 10% of the cement was replaced with used paint. The effect of this replacement is dramatic especially on the behavior of the mortar at relatively short times after preparation. It is observed that the distribution of T_2 relaxation time measured at 3 days after preparation shows approximately four peaks, two of which correspond to bound water and probably to small or ultra small pores [8-10]. They are superimposed. The droplet corresponding to the water in these small pores is observed as a shoulder over the peak related to the bound water.

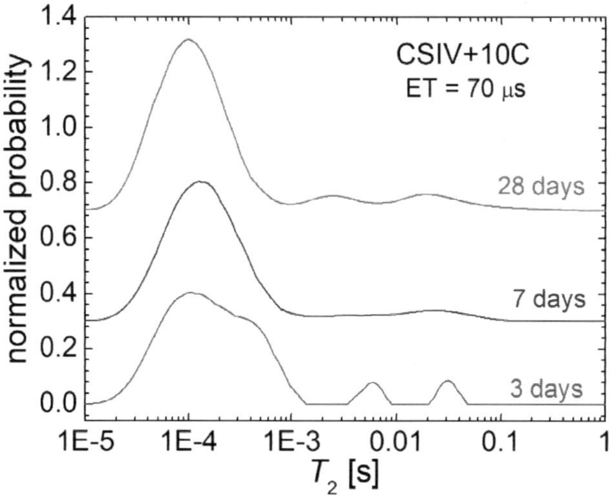

Figure 2. Normalized distribution of T_2 for the CSIV 10C mortar recipe.

By hardening the CSIV 10C plaster mortar at 7 days and at 28 days, we do not notice a significant shift of the distribution center corresponding to the connected water, instead the distributions corresponding to the water in the pores change quite a lot [11,12]. Thus, at 7 days the CSIV 10C mortar for which 10% of the cement was replaced with used paint has almost no pores, and at 28 days some structures begin to appear that can be filled with water and that can be considered as small pores and medium pores, but their relative amplitude is much smaller than in the case of CSIV plaster mortar. Therefore, we can say that the replacement of 10% of cement with used paint could have an effect of destroying these pores which implicitly leads to increased compactness of the material, increased mechanical strengths and increased impermeability.

2.3. NMR investigations on the CSIV 20P plaster mortar recipe

Figure 3 shows the T_2 relaxation time distributions for CSIV 20P plaster mortar for which 20% of the amount of water was replaced with used paint. Here, too, the effect is quite significant, so at 3 days a distribution of T_2 is obtained which has 3-4 peaks, but 2 of them are located at quite low values of T_2, and can be superimposed, so those corresponding to the water bound and water from very small pores. Also in this case, 2 peaks are obtained corresponding to the water from small to medium pores and medium pores. As with CSIV plaster mortar, the distribution of T_2 measured for CSIV 20P mortar by replacing 20% of the amount of water with used paint, the distributions are similar, so we have the highest amount of hydrogen located in hydrating water and small amounts of hydrogen in small to medium and medium pores.

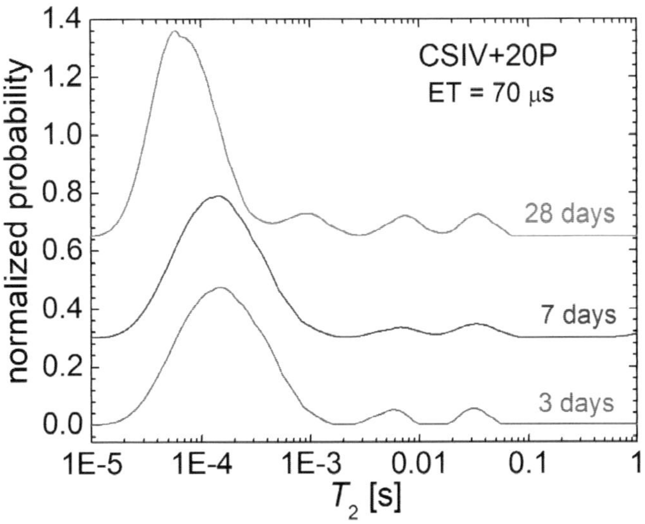

Figure 3. Normalized distribution of T_2 for the CSIV 20P mortar recipe.

Instead, after 28 days of preparation by replacing 20% of the amount of water with used paint, a distribution of T_2 is obtained which has four clear peaks, one of them corresponding to the hydration water and for which we have a hydrogen strongly bound to the rigid components, which indicates that a large amount of this hydrogen is found in the sample studied. But we can also see approximately equal amounts of hydrogen that are located in small pores, small to medium pores and medium pores.

3. Interpretation of NMR results

In order to better characterize the effect of replacing water and cement with used paint, below in Figure 4, Figure 5 and Figure 6 are presented the distributions of the relaxation time T_2 for the three samples: CSIV plaster mortar, CSIV 10C plaster mortar for which replaced 10% of the cement with used paint and CSIV 20P plaster mortar for which 20% of the water was replaced with used paint at 3 days, 7 days and 28 days.

3.1. NMR results at 3 days

Thus, Figure 4 shows the distributions for T_2 measured for samples 3 days after preparation and significant changes are observed both in the area of the peaks corresponding to the hydration water and in the area corresponding to the peaks associated with the hydrogen in the water found in small to medium and large pores. We can see a similarity between CSIV plaster mortar and CSIV 20P plaster mortar for which 20% of the water was replaced with used paint for the hydration water drop, but CSIV 10C by replacing 10% of the cement with used paint leads to an increase in the inhomogeneity of the material, in fact the rigid components of this material at least 3 days after preparation. In contrast, CSIV 10C mortar for which 10% of the cement was replaced with used paint has a clearer to medium to medium pore structure, which is clearer from the narrower width of the peaks between 1 millisecond and 100 milliseconds corresponding to these small to medium and medium pores. After that, it is observed that the replacement of 20% of water with used paint in CSIV 20P plaster mortar leads to an increase in the inhomogeneities of the porous structure 3 days after preparation, and the CSIV plaster mortar shows the highest inhomogeneity of the structures from which small to averages and averages pores will later form. This is observed by the fact that the peaks corresponding to the two types of pores are superimposed and are no longer distinct as in the case of the other 2 types of mortars CSIV 10C and CSIV 20P.

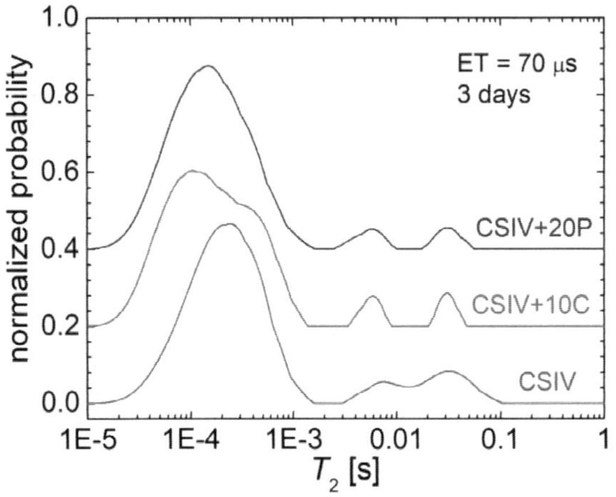

Figure 4. Normalized distribution of T_2 at 3 days.

3.2. NMR results at 7 days

On the other hand, 7 days after the preparation of the plaster mortars, it seems that it is going through an intermediate phase quite similar for all three plaster mortars, Figure 5. Again, it appears that the T_2 distributions measured for CSIV plaster mortar and CSIV 20P mortar for which 20% of the amount of water was replaced with used paint are quite similar, with small to medium pores beginning to form in the case of mortar. CSIV 20P with 20% replacement of water with used paint are better pronounced. Otherwise, it seems that the peaks corresponding to the bound water for all three types of pores are quite similar, which indicates that in this area of the hydrogen bound to the components of the mortar are quite similar.

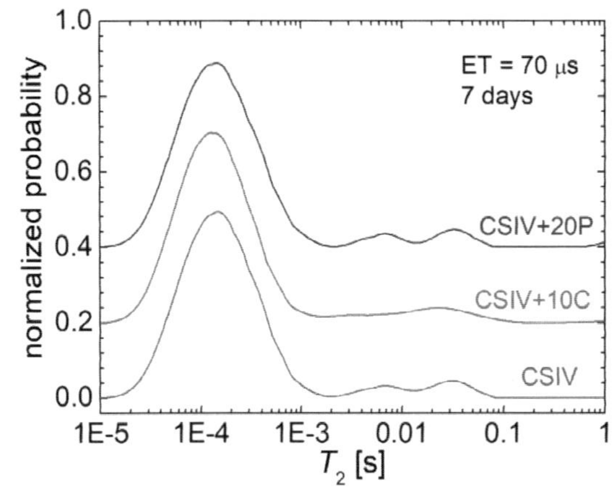

Figure 5. Normalized distribution of T_2 at 7 days.

Replacing 10% of the cement with used paint in CSIV 10C plaster mortars recipe leads us to the fact that at 7 days this pore structure does not really exist, and if we compare it with the distribution from 3 days it indicates that it has somehow formed, but it is not a final formation, which in time begins to destroy itself.

3.3. NMR results at 28 days

If we characterize the samples from the point of view of relaxometry, 28 days after preparation, we can see the formation of these types of small to medium and medium pores, the most pronounced CSIV plaster mortar, followed by CSIV 20P plaster mortar for which have replaced 20% of the amount of water with used paint and for which we can see the following two elements: the drop corresponding to the bound water is at a value of T_2, average value or maximum value, lower than in the other two cases, which it indicates that hydrogen is found in more rigid components, it could also lead to a rigidity of the whole material. In addition, there is the appearance of a type of pores but very small in size, which is not the case with the other two samples. The effect of introducing the used paint into the CSIV 10C plaster mortar recipe, by replacing 10% of the amount of cement with the used paint, leads to the destruction of the porous structure. As a result, the material becomes more compact, the mechanical strength increases and its impermeability is increased, Figure 6.

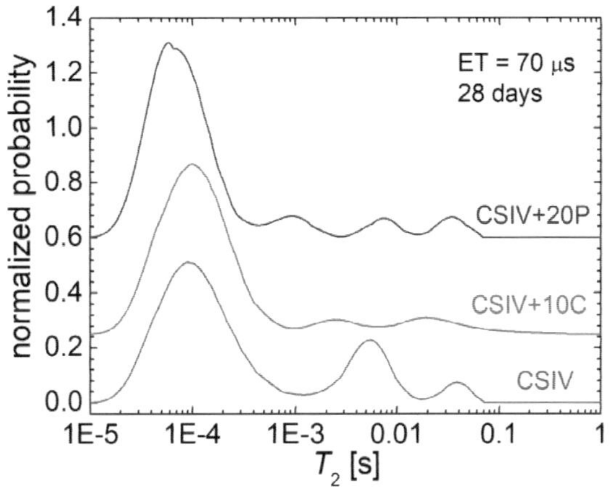

Figure 6. Normalized distribution of T_2 at 28 days.

At the same time, in the case of CSIV 10C it is observed that the water that remains at 28 days after preparation in small to medium and medium pores is found in a rather small amount, which indicates that their number is quite small or this water. it has evaporated anyway, and the fact that they are quite wide indicates an increase in heterogeneities, so this small pore can barely be distinguished from the medium ones.

4. Conclusions

The intervention of used paint applied to the concept of sustainable development makes the material more compact, respectively more waterproof.

Used paint adds value to the CSIV 10C mortar recipe in terms of compactness, impermeability and color. This way, the beneficiaries no longer have to use waterproof or dyeing solutions to add color to the plaster mortar.

By using used paint, the amount of cement is reduced, resulting in a cheaper material.

Acknowledgement

This paper was financially supported by the Project "Entrepreneurial competences and excellence research in doctoral and postdoctoral programs - ANTREDOC", project co-funded by the European Social Fund financing agreement no. 56437/24.07.2019.

References

[1] Siomin A C, Țințișan M L, Zaharie A and Manea D L 2022 Recovery of Used Paints in the Field of Plaster Mortars, *Springer - Lecture Notes in Networks and Systems* **386** 129-138
[2] United Nations 2015 The 2030 Agenda for Sustainable Development
[3] Tomislav K 2018 *Zagreb International Review of Economics and Business* **21**(1) 67-94
[4] Fechete R, Demco D, Dumitrita Moldovan C, Chelcea I and Culea E 2010 Rezonanța Magnetică Nucleară: Metode clasice și moderne (Cluj Napoca: Risoprint)

[5] Ciubară A, Burlea SL, Axinte M, Cimpoeşu R, Chicet DL, Manole V, Burlea G and Cimpoeşu N 2018 *IOP Conference Series: Materials Science and Engineering* **444(3)** 032013.

[6] Axinte M, Cimpoeşu R, Chicet DL, Manole V, Cimpoesu N, Burlea SL, Ciubară A and Burlea G 2019 *5th International Conference on Control, Automation and Robotics (ICCAR)* 812-816. doi: 10.1109/ICCAR.2019.8813696.

[7] Baltatu MS, Vizureanu P, Cimpoesu R, Abdullah MMA and Sandu AV 2016 *Revista de chimie* **67(10)** 2100-2102.

[8] Cimpoesu N, Stanciu S, Vizureanu P, Cimpoesu R, Achitei DC and Ionita I 2014 Journal of mining and metallurgy section B-metallurgy **50(1)** 69-76. DOI: 10.2298/JMMB121206010C.

[9] Burduhos-Nergis D-P, Sandu A-V, Burduhos-Nergis D-D, Darabont D-C, Comaneci R-I and Bejinariu C 2019 *MATEC Web of Conferences* **290** 12004

[10] Abtew MA, Boussu F, Bruniaux P, Loghin C, Cristian I, Chen Y and Wang L 2018 *AIP Conference Proceedings* **1960** 020001. doi.org/10.1063/1.5034802.

[11] Yahya Z, Abdullah MMAB, Ramli NM, Burduhos-Nergis DD and Abd Razak R 2018 *IOP Conf. Ser.: Mater. Sci. Eng.* **374** 012068. DOI: 10.1088/1757-899X/374/1/012068..

[12] Nergis, DDB, Vizureanu P, Sandu AV, Nergis, DPB and Bejinariu C 2022 *Materials* **15(1)** 202. DOI: 10.3390/ma15010202.

Research on valorisation of spent garnets as addition in cementitious materials – preliminary experimental evaluation

C Baeră[1,2,3], R Chendeş[1], A Gruin[3], A Perianu[4], V Vasile[5] and L Varga[1]

[1]Politehnica University of Timisoara, Timisoara, Romania
[2]IOSUD-UTCN Doctoral School, Cluj-Napoca, Romania
[3]NIRD URBAN-INCERC Timisoara Branch, Romania
[4]ISIM Timisoara, Romania
[5]NIRD URBAN-INCERC Bucharest Branch, Romania

E-mail: cornelia.baera@upt.ro, cornelia.baera@incd.ro

Abstract. Abrasive waterjet (AWJ) material processing represents a relatively new and extremely efficient method in the specific industries, implying various applicability areas, attributed to different materials, with different properties and domains of use. Besides numerous advantages of the AWJ cutting and generally of the material processing techniques, they also involve the waste generation: the abrasive sands are converted into sludge material, collected into recipients and further on, after natural drying, they become waste dumps randomly deserted. The enlargement of the AWJ techniques in the latest years determines the corresponding increase of the associated Garnet wastes (Spent Garnets, SG), thus leading towards the clear need of identifying opportunities for their recycling and valorisation. Aggregates are basic raw materials in the production of concrete, mortar and plasters, composite cementitious and/or cement-free materials (geopolymer concrete) in the building materials and generally, in the construction industry. The rapid growth of the population, recorded mainly in urban areas, determines an increasing demand on housing facilities and consequently, on concrete production and aggregates consumption. Aggregates and sand, mainly exploited from riverbeds or quarries, represent exhaustible natural resource for which substitution solutions need to be found, in order to control and reduce their extraction from the natural landscape. Considering the superposition of this independent cause and effect situations, the rapid growth of Spent Garnets (SG) landfills, generated by AWJ processes, and, on the other hand, the need to substitute the natural aggregate in construction industry, a reliable solution could emerge from the potential valorisation of SG wastes as partial or even complete substitution of sands / aggregates in the composition of construction materials. The present paper offers a preliminary overview regarding the possibility of incorporating SG wastes of local production in usual cement-based materials, as partial substitute of the aggregate, for the double purpose of waste management implementation and natural resources protection, on the transition path towards the implementing the Circular Economy (CE) concept in the Romanian industry.

1. Introduction

The acute state of general environmental degradation dominates the current global context, due to already known factors like change of climate, global warming, poor recycling efficiency, fast increase of greenhouse gas emissions, mainly generated by the industry trying to fulfil the humanity's everyday needs and activity, in superposition with high consumption of natural resources, etc. Another important factor is represented by the exponential growth of population, recorded mainly around the existing urban centres with already high population density, which leads to the corresponding need increase for housing and complementary infrastructure facilities [1]. Globally, the population growth is estimated from 2 to 4 billion people by 2030; as consequence, the construction industry and the building materials sector are expected to record a huge expansion as well, with high demands in terms of raw materials and energy consumption. In the near future it is expected that the construction production to exceed that of the last 4000 years [2].

Aggregates, sand included, are critical raw materials in concrete and mortar mixes, quantitatively and qualitatively as well: they have a large share in concrete mass with respect to the other constituents and they also influence considerably its strength and also its durability performance. Unfortunately, they represent natural, exhaustible resource, extracted from riverbeds or quarries. Their uncontrolled and excessive exploitation, at European level but especially nationally, generates negative environmental impact, Figure 1, with already recorded devastating consequences: erosion of shorelines, low groundwater level, local fauna and flora damage, landslides, accidents and drowning, etc. [3, 4]. The Romanian legislation is poorly adapted to the nowadays context, permitting its elution and even abuse related to landscape aggressive exploitation, with long term negative effects. Supplementary, it was noticed that Romanian aggregate resources are lacking sand and generally the fine grain part. Consequently, proper solutions need to be identified to balance these negative aspects.

Figure 1. Environmental degradation due to abusive aggregate exploitation [3,4].

Concrete and cement-based materials, globally considered as most used building materials, are already identified suitable to incorporate diverse residues or industrial by-products, as additions or raw material substitution in their mass, with good performance in terms of physical, mechanical and also durability. Plastic and ceramic, glass waste, fly ash and slag, bottom ash and limestone slurry, etc., are already valorised in mortar or concrete mixes, due to intense research in the topic, without compromising the initial characteristics of the matrix [5-12]. This way, green directions for recycling purposes are already established, simultaneously solving two independent problems: exhaustion of natural resource and waste management considering the Circular Economy (CE) concept, namely the innovative waste conversion towards new life cycles.

Garnet type wastes result from the material processing techniques based on water jet and additional abrasive materials, usually Garnets (abrasive waterjet, AWJ process). Their use and applicability are constantly growing around the world, Romania included; the resulted waste, Spent Garnets (SGs) is also increasing. SG waste is randomly stored in landfills, producing mutation of the landscape and pollution. Recent studies, which grew in number and consistency in the latest years, evaluate the GS potential as addition in concrete mixes, with preliminary encouraging results [5-12]. In this general context the current research is focused on valorising the SG locally produced as partial substitute of ordinary sand in mortar and concrete mixes, with respect to the CE principles.

2. Abrasive waterjet (AWJ) methods for material industrial processing

AWJ material processing represents a relatively new and extremely efficient method in the specific industries, implying various applicability areas, attributed to different materials, with different properties and domains of use, Figure 2. The material processing, besides cutting includes drilling, milling, sandblasting and cleaning as well, as the versatility of the technique and the associated research for continuous improvement determines its large applicability in many complementary or distinct fields [13-16].

Figure 2. Waterjet cutting operations with complementary Garnet abrasive – ISIM Timisoara [15].

The use of waterjets is not new: the Egyptians used it also combined with sand for mining and cleaning purposes; when the technology allowed the increase of water pressure, the technique was adapted for forestry activities, as well. The abrasive inclusion in the process improved the technique and led to better and faster results and also determined the expansion of use: thicker sections of harder material (metal included) were able to be cut, with improved precisions of shapes and edges. Cleaning and paint removal techniques in the maritime industry and ship maintenance started to use the high-pressure sandblasters with pressurized steam. Nowadays, Aluminium, Alloys, Stainless Steel and generally metal products are generated by these methods.

3. Materials and methods

3.1. General approach of the experimental study

The major goal of this experimental study is to evaluate the potential of SG wastes, locally produced, as partial replacement of sand in a regular cement-based mortar. The research protocol is simple and in accordance to previous study in the topic [17-21]: a reference mix was proposed (R), followed by the relevant parametric variation, namely the sand partial substitution with SG wastes of two local sources (SGI, SGII). The natural, regular sand substitution was performed by weight and it ranged from 10% to 50%, with respect to the reference, in accordance to previous studies' classical approach. The raw material analysis was also performed, followed by the evaluation of the mortar mix, fresh and hardened state (early age characteristics).

3.2. Raw materials

The raw materials used for the mortar mixes are of local sources, the Reference mix (no substitution) and also the derived mixes (SG mixes) were produced using the following constituents:

- Portland Cement, CEM II/A-LL 42.5 R (C), produced by Carpatcement (Figure 3, a);
- Natural aggregates of granular class 0/4 (sand, S), (Figure 3, b);
- Water (tap water);
- Spent Garnet sand (SG), waste collected from two distinct local sources, SG I (Figure 3, c) and SG II (Figure 3, d).

No additives were used in this preliminary compositional design approach.

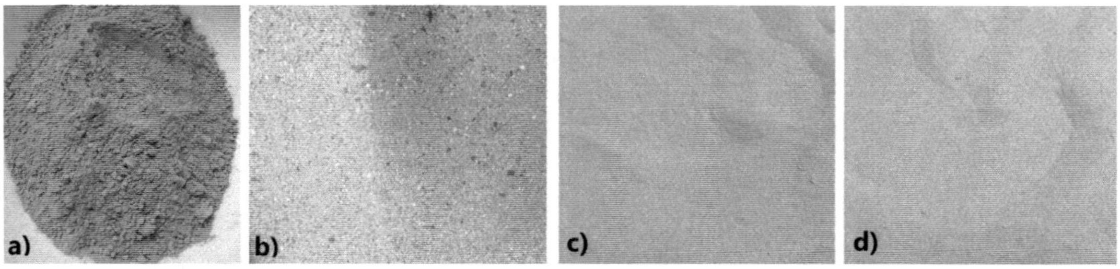

Figure 3. Raw materials for mortar mix design: a) Binding system - Portland Cement; Aggregates: b) Natural sand (0/4); c); SG I; d) SG II.

The raw Garnet material, used in the AWJ cutting installation which generated the SG I, represents an 80 Mesh Garnet, in accordance with the distributor specifications [23]. It is also specified that it is an ecologic material, in accordance to the European legislation, namely the Regulation (EC) No 1272/2008 and Regulation (EC) No 1907/2006, producing no environmental damage and being derived from the natural almond ore mined, further subjected to several processing operations to obtain a high-quality abrasive material. Still, the associated technical sheet does not contain any clear information regarding the potential toxicity of the Garnet abrasive. Specific studies mentioned by the literature stated that the virgin Garnet material generally falls within the acceptance limit regarding the heavy metals content (Arsenic (As), Barium (Ba), Cadmium (Cd), Chromium (Cr), Lead (Pb), Selenium (Se), Silver (Ag), Zinc (Zn) and Copper (Cu) [24-27]. Obviously, the Garnet wastes (SG) heavy metal content is influenced by the processed material subjected to AWJ cutting. Therefore, leaching analysis of the resulted Garnet wastes could represent a critical evaluation for their future integration in cement-based building materials.

The preliminary evaluation of the aggregates used for the experimental study, namely the natural sand 0/4 and the SG samples, SG I and SG II, considered for the 0/4 sand partial substitution, was performed in terms of relevant physical and chemical characteristics.

The aggregate grading is derived from sieve analysis, performed on all samples: S, SG I and SG II, is presented in Figure 4.

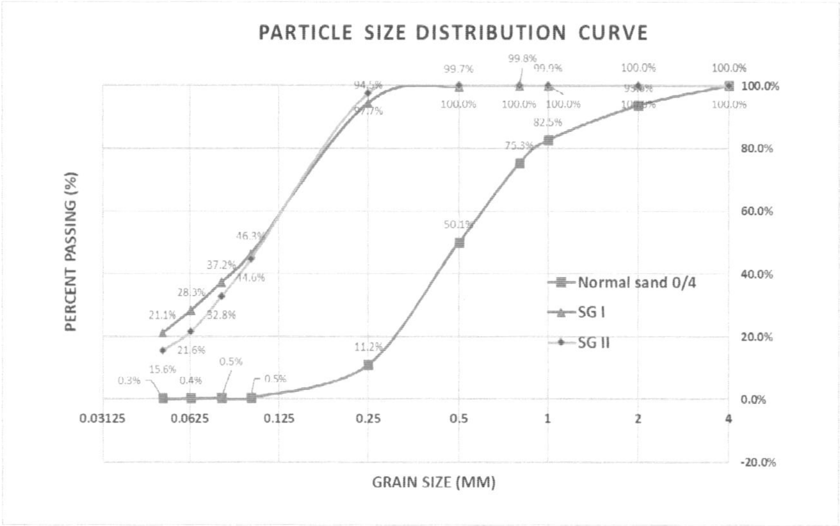

Figure 4. Raw materials for mortar mix design: a) Binding system- Portland Cement; Aggregates: b) Natural sand (0/4); c); SG I; d) SG II.

As expected, the natural sand 0/4 (S) is lacking the fine grain content, as the corresponding grading curve is presenting; the representative fractions are ranging between 0.100 mm and 1 mm sieve or, presented as cumulative retained mass percent, as follows: 1/2 mm: 17.5%; 0.8/1: 24.75%; 0.5/0.8: 49.9%; 0.25/0.5: 88.85%; 0.1/0.25: 99.45%. The two SG samples present a common granular distribution pattern, with smaller grain content for SG II, but generally a distinctive pattern with respect with the S sample: the S sieve content is ranging basically from 2 mm to 0.25 (0.1) mm sieve and the SGs range from 0.1 mm sieve to the pan (receiver). This analysis indicates the substitution potential of the SG material, for a further optimised grain size distribution of a blend aggregate, S+SG, with direct use in the design of green cementitious materials.

The physical characteristics of the aggregates, currently determined, are presented in Table 1, future analysis being also considered, for a proper comparative evaluation. The higher bulk density of the SG samples is considered to be due to the increased content of iron oxide (Fe_2O_3), as the preliminary chemical analysis indicates (Table 2).

Table 1. Physical properties of aggregates.

Composition code	Natural Sand 0/4 (S)	Spent garnet I (SG I)	Spent garnet II (SG II)
Bulk density, (kg/m^3)	1550	2000	2120
Dry net density, (kg/m^3)	-	2690	-
Water absorption (WA, %)	-	2.4	-
Fineness modulus	6.86	3.74	3.88

The chemical composition of SG I material was determined by specific XRF analysis, the results are presented in Table 2.

Table 2. SG I chemical composition (XRF analysis).

Oxides	**SG I, %**	**Raw Garnet, % [23]**
SiO_2	36.65	35.20
Al_2O_3	17.54	18.90
Fe_2O_3	28.58	38.10
CaO	6.27	
MgO	5.88	2.50
SO_3	0.05	
Na_2O	0.22	0.03
K_2O	1.07	0.02
P_2O_5	0.08	0.06
TiO_2	2.12	2.00
Cr_2O_3	0.05	
Mn_2O_3	1.37	
ZnO	0.26	
SrO	0.00	
ZrO_2		0.03
L.O.I.	0.00	

3.3. Mortar mix design and mixing procedure

The eco-mortar mix design, with incremental increase of SG content as regular sand substitution, was developed in accordance to the scientific literature, which lately showed an increased interest towards the topic. Two Reference mortars were produced, R1 and R2, with basic raw materials: regular sand

0/4, cement and water (different water content). The initial sand replacement considered the S 25% and 50% substitution with SG material, but some unexpected demoulding problems with 50% sand substitution mortar changed the approach towards a moderate incremental S substitution: 10%, 20% and also 30%.

Table 3 presents the mix proportions of the references, R1 and R2 and SG mixes as well, expressed in terms of the specific ratios with respect to the cement (C) content (C = 1.0).

Table 3. Mix proportions of eco-mortars, Reference and SGs substitution.

Components / Mix design	C	S 0/4	SG I	SG II	W/C	A/C
R1	1	3	-	-	0.61	3.00
R2	1	3	-	-	0.56	3.00
SG I, 25%	1	2.25	0.75	-	0.56	3.00
SG II, 50%	1	1.50	1.50	-	0.56	3.00
SG I, 10%	1	2.70	0.30	-	0.56	3.00
SG I, 20%	1	2.40	0.60	-	0.56	3.00
SG I, 30%	1	2.10	0.90	-	0.56	3.00
SG II, 10%	1	2.70	-	0.30	0.56	3.00
SG II, 20%	1	2.40	-	0.60	0.56	3.00
SG II, 30%	1	2.10	-	0.90	0.56	3.00

The mixing sequences were established considering the typical laboratory mixture suitable for low volume batches, in accordance to the EN 196-1 [19] specifications:

- Water (W) and Cement (C) initial addition to the mixing vessel, W +C;
- 30 seconds of low-speed mixing of W and C;
- Aggregate addition during 30 s of low-speed mixing;
- 30 seconds of fast-speed mixing;
- 90 seconds of rest, while bowl scrapping in the first 30 seconds of the recommended break, for better mixture homogenizing;
- 60 seconds of fast-speed mixing.

After mixing, the fresh state characteristics of the materials was evaluated and the mortars were cast into prismatic, metallic moulds, and placed for initial 24 h curing at the temperature T (20±1)°C and relative humidity RH (90±5)%. After 24 hours, the prismatic specimens were removed from the moulds and cured by water immersion at the temperature T (20±1)°C, until the early age testing, performed at 7 days.

4. Results and discussions

The fresh state characteristics of the green mortars, developed by specific substitution of the regular sand with Garnet wastes are recorded, together with early age hardened state properties, considered to be critical for the future development of the study.

The initial Reference mortar, M1, was designed in accordance to a consistency of around 10 cm (determination by plunger penetration, in accordance to EN 1015-4), which lead to a specific ratio of W/C of 0.61 (Table 3). The first substitution mortars required less water, so Reference M2 was considered, with a reduced C/W ratio, which was kept constant for this stage of experimental procedure.

4.1. Fresh state aspect

All mixtures presented a homogeneous fresh state composition, specific to plain, regular mortars. The regular sand 0/4 substitution with SG wastes led to an improved cohesivity of the mixtures, as due to fine particle increased content of the SG, as expected (Figure 5). The cohesivity of mixes and the creamy aspect of the mortars increases as the substitution percent increases. The fresh state aspect of all developed mixes, reference and eco-mortars, is mostly similar, with no dramatic changes from one batch to another, considering the W/C constant at 0.56 (except for the R1).

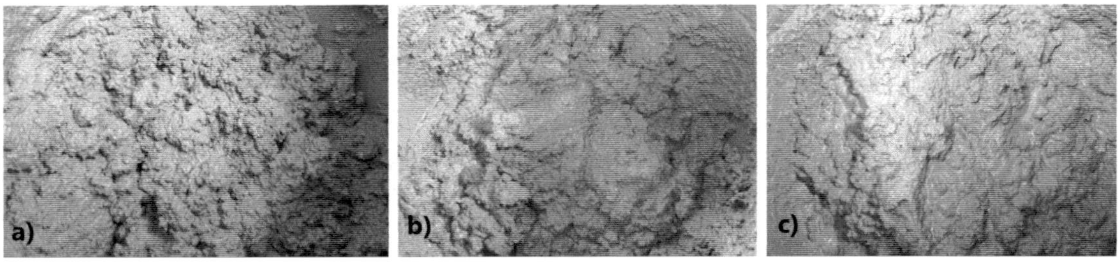

Figure 5. Fresh state aspect of the mixes: a) R2 mix; b) SG I, 25% mix; b) SG II, 20% mix.

4.2. Consistency of the mortar mixes

The mortar consistency (Figure 6) is determined by the means of the standard cone method [28]. The standard cone (Figure 6, a) put in the vertical position with its peak centred on the fresh mortar testing sample (Figure 6, b), is released and it penetrates the mortar mass by the means of its own weight. The penetration depth is read on the cone, in cm. The method provides a relevant comparative analysis regarding the workability of the material. The results are recorded in Table 4 and they emphasise the beneficial contribution that the SG wastes can provide in the eco type mortars.

Figure 6. Fresh state aspect of the mixes: a) R2 mix; b) SG I, 25% mix; b) SG II, 20% mix.

4.3. Hardened state densities

The volumetric mass recorded immediately after specimens' removal from the moulds, 24 h after mixing and casting, is presented in Table 5, together with the 7-day saturated apparent density, determined in accordance to EN 12390-7 [29]. The results indicate an obvious density increase pattern, which could be directly related to the replacement percentage of the regular sand with SG wastes, suggesting an improved matrix packaging when substitution percentage increases; in the same time, the SG II mortars seem to develop a denser mix when compared to the SG I. The dry state density, the most relevant characteristic in terms of densities, performed in accordance to the reference method specified by EN 1015-10 [30], is on-going testing and it is expected to confirm the already identified increase pattern.

Table 4. Consistency of eco-mortars, Reference and SGs substitution specimens.

Mix design	**Consistency (mean value), cm**
R1	8.5
R2	4.5
SG I, 25%	6
SG II, 50%	5
SG I, 10%	8
SG I, 20%	7
SG I, 30%	6
SG II, 10%	6
SG II, 20%	8
SG II, 30%	7

Table 5. Hardened state densities, Reference and SGs substitution specimens.

Mix design	**Volumetric mass (after demoulding), kg/m^3**	**Saturated apparent density, kg/m^3**
R1	2180	2200
R2	2160	2180
SG I, 25%	2360	2370
SG II, 50%	2440	2390
SG I, 10%	2210	2230
SG I, 20%	2320	2330
SG I, 30%	2360	2370
SG II, 10%	2270	2290
SG II, 20%	2310	2330
SG II, 30%	2400	2420

4.4. Early age mechanical properties

The flexural and compressive strengths, critical mechanical characteristics for the mix evaluation, were performed at early age, namely 7 days, by using the EN 196-1 [31] and EN 1015-11 [32] methods. Prismatic specimens, 40 x 40 x 160 mm^3, removed from the moulds 24h after casting, were cured by water immersion at T (21 ± 2) °C until testing. The results are presented in Table 6, namely the early age, 7-day flexural strength and also the 7-day compressive strength, together with the corresponding strength increases/decreases, evaluated with respect to the relevant R2 reference mix. The graphical representation of the results is presented in Figure 6 and Figure 7.

Table 6. Early age mechanical performance: flexural and compressive strength.

Mix design	3PB Flexural resistance		Compression resistance	
	Strength (MPa)	Strength gain/loss vs R2 (%)	Strength (MPa)	Strength gain/loss vs R2 (%)
R1	5.9	-15.7	22.0	-7.8
R2	***7.0***	-	***23.8***	-
SG I, 25%	7.6	8.6	27.7	16.2
SG II, 50%	7.6	8.6	33.0	38.5
SG I, 10%	6.6	-5.7	24.2	1.4
SG I, 20%	7.5	7.1	26.0	9.0
SG I, 30%	7.9	12.9	29.5	23.9
SG II, 10%	7.5	7.1	25.5	7.0
SG II, 20%	7.5	7.1	25.7	7.7
SG II, 30%	8.2	17.1	28.3	18.5

a)

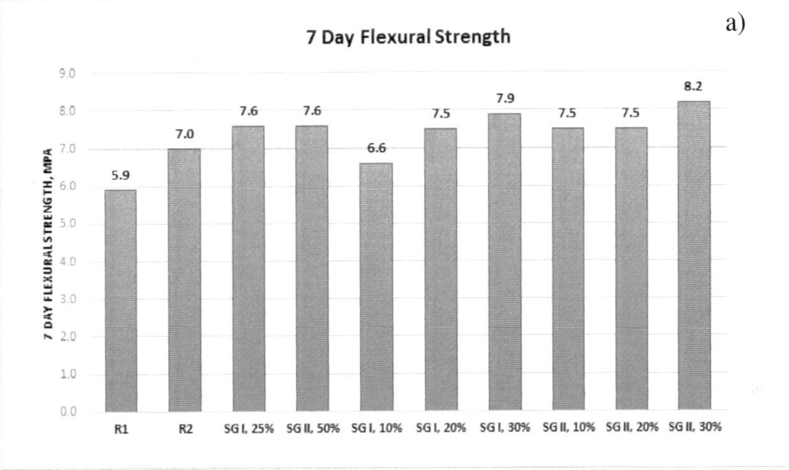

b)

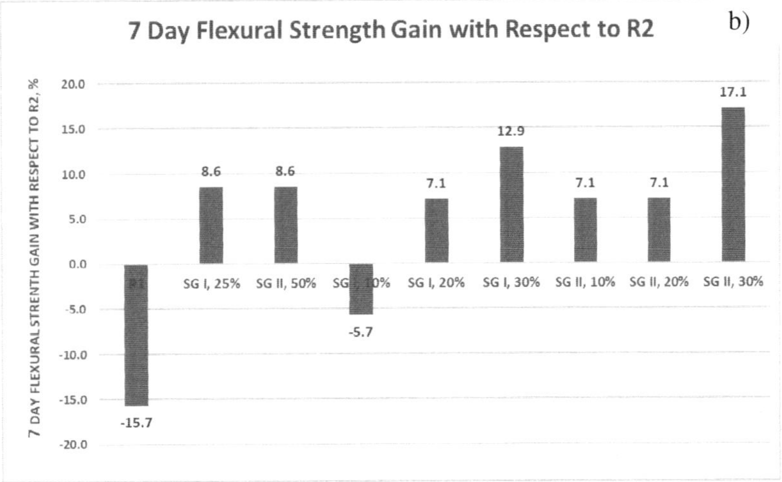

Figure 7. Early age mechanical strength: a) 7-day flexural strength; b) 7-day flexural strength gain/loss, with respect to R2.

The results are confirming the increase pattern identified for density evaluation, the mortar performance increases with respect to the substitution percent grows.

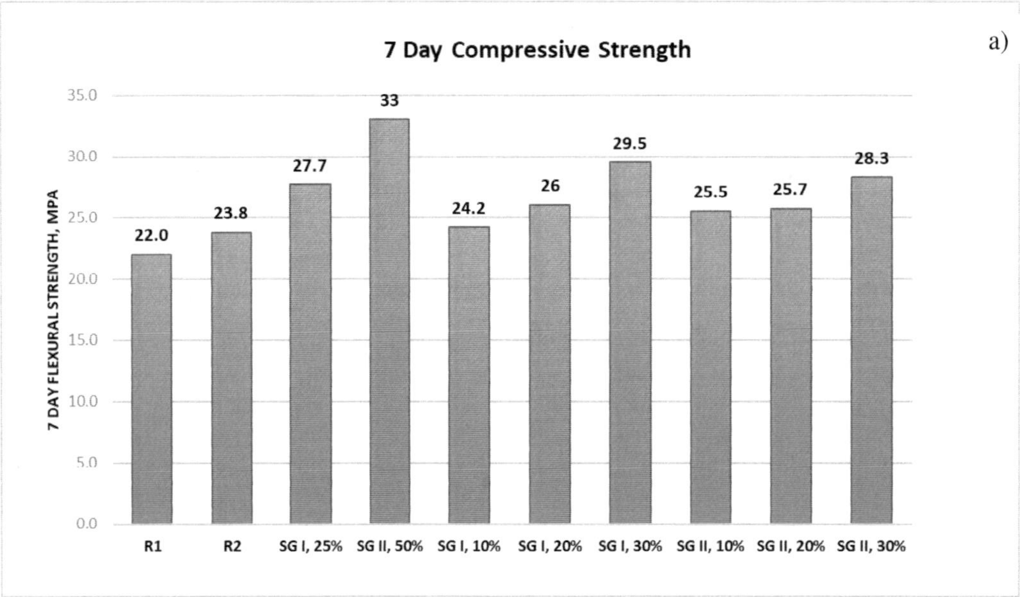

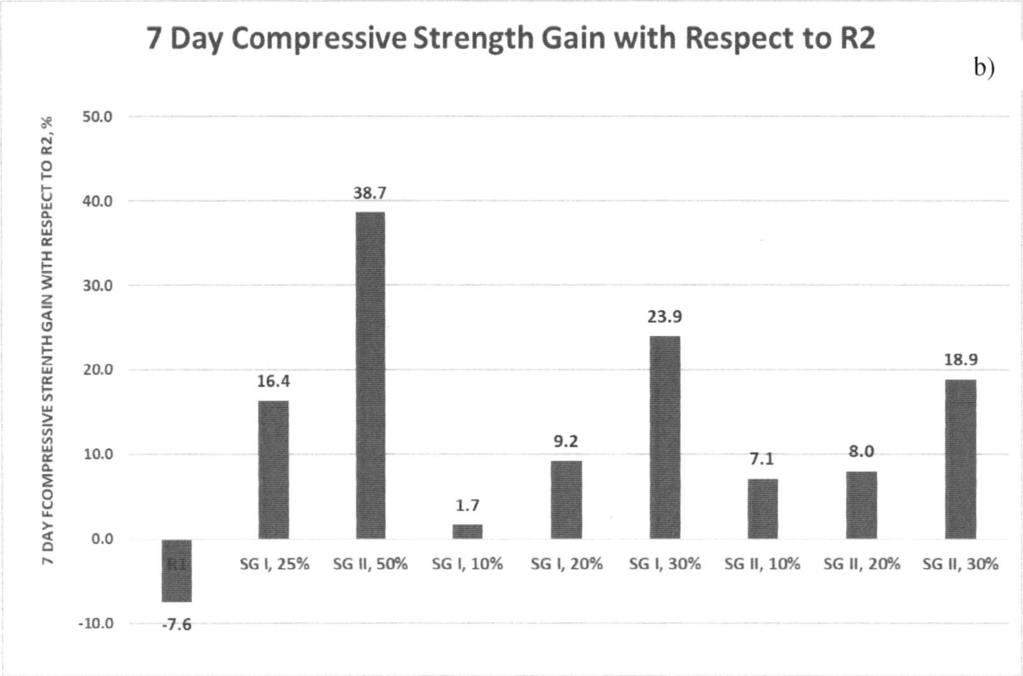

Figure 8. Early age mechanical strength: a) 7-day compression strength; b) 7-day compression strength gain/loss, with respect to R2.

5. Conclusions and future perspectives

The preliminary testing and the corresponding results are offering encouraging conclusions regarding the viability of the concept. The SG wastes can be positively integrated in cement-based materials, as they prove themselves compatible to the binding, cementitious system and also to the regular sand. The initial physical and chemical testing of the SG as potential substitute for regular 0/4 local sand, together with physical and mechanical performance of the initial proposed eco-mortars obtained in this preliminary experimental investigation are confirming, partially, the conclusions of recent studies in the topic and also the potential of abrasive Garnet wastes for integration in cement-based building materials.

The common pattern of the samples developed with two distinct SG wastes, resulted from two distinct local producers, are confirming the accessibility of identifying the optimising mix design approach for the transfer to the industry and commercial use. In the same time, relevant mechanical performance and durability characteristics of the derived cementitious composites, together with specific leaching tests of the SG derived from several sources are mandatory identification procedures in order to correctly identify the applicability directory lines for the Garnet wastes valorisation in the construction industry.

Acknowledgments

This paper is supported by the Project "Entrepreneurial competences and excellence research in doctoral and postdoctoral programs - ANTREDOC", project co-funded by the European Social Fund financing agreement no. 56437/24.07.2019 and also by the Programme: Research for sustainable and ecological integrated solutions for space development and safety of the built environment, with advanced potential for open innovation – "ECOSMARTCONS", Programme code: PN 19 33 04 02: "Sustainable solutions for ensuring the population health and safety within the concept of open innovation and environmental preservation", financed by the Romanian Government.

References

[1] Kharas H *The unprecedented expansion of the global middle class an update* 2017 (https://think-asia.org/bitstream/handle/11540/7251/global_20170228_global-middle-class.pdf?sequence=1)
[2] Eberhardt L C M, Birgisdottir H and Birkved M 2019 IOP Conference Series: Materials Science and Engineering, **471**, No. 9, p. 092051, 2019
[3] https://ecointeligenta.ro/societate-amendata-cu-30-000-de-lei-pentru-exploatarea-de-agregate-fara-autorizatie-de-mediu/
[4] https://www.acdb.ro/interzise-prin-lege-si-totusi-practicate-legal.html
[5] Baeră C 2016 *Self-Healing capacity of Advanced Engineered Cementitious Composites*, (Romania: Technical University of Cluj-Napoca)
[6] Baeră C, Szilagyi H Lăzărescu A and Mircea A M 2019 *Proc. fib Symp. 2019: Concr. - Innov. Mater. Des. Struct.* 147831
[7] Corbu O, Ioani A M, Al Bakri M M A, Meiţă V, Szilagyi H and Sandu A V 2015 *Key Eng. Mater.* **660** 237–43
[8] Popovici A, Corbu O, Popita G E, Rosu C, Proorocu M., Sandu A V and Abdullah M M A B 2015 *Mater Plast* **52**(4) 588–592
[9] Burduhos Nergis D D, Vizureanu P, Ardelean I, Sandu A V, Corbu O C and Matei E 2020. *Materials* **13**(14) 3211
[10] Baeră C, Szilagyi H, Matei C, Hegyi A, Lăzărescu A and Mircea A C 2019 *MATEC Web Conf.* 01001
[11] Lăzărescu A, Baeră C and Szilagyi H 2019 *Proc. EUROCOALASH* 145–156
[12] Szilagyi H, Baeră C, Lăzărescu A and Mircea A C, 2018 *Constructii*, **19**(1/2) 3–10
[13] Perianu I A, Mnerie G V, Cojocaru R and Binchiciu E F 2021 *Key Eng. Mater.* **890** 147–151
[14] Perianu I A, Mitelea I and Şerban V A 2014 *Adv. Mat. Res.* **1029** 176–181
[15] Baeră C, Vasile V, Matei C, Gruin A, Szilagyi H and Perianu I A 2021 *Adv. Mat. Res.* **1164** 87–96
[16] Perianu I A, Ionescu D and Verbitchi V 2017 *Welding and Material Testing* **1**

[17] Muttashar H L, Hussin M W, Ariffin M A M, Mirza J, Hasanah N and Shettima A U 2017 *J. Teknol.* **79**(3)
[18] Muttashar H L, Mohd Ariffin M A, Hussin M W and Ishaq S B 2018 *Mag. Concr. Res.* **70**(11) 558–569
[19] Budiea A M A, Sek W Z, Mokhatar S N, Muthusamy K and Yusoff A R M 2021 *IOP Conf. Ser.: Mater. Sci. Eng.* **1144**(1) 012018
[20] Kadir M A Ab, Khiyon M I, Sam A R M, Kueh A B H, Lim N H A S, Mastor M N M A, Zuhan N and Mohamed R N 2019 *J. Struct. Fire Eng.* **10**(4) (2019)
[21] Kanta N R and Ponnada M R 2021 *World J. Eng.*
[22] https://gritsablare.ro/granat-rosu-garnet-debitare.html
[23] Jamaludin N F A, Muthusamy K, Isa N N, Jaafar M F M and Ghazali N, 2021 *Proc. Materials Today*
[24] Muttashar H L, Ali N B, Ariffin M A M and Hussin M W 2018 *Case Stud. Constr. Mater.* **8** 87–96
[25] Huseien G F, Sam A R M, Shah K W, Budiea A M A and Mirza J, 2019 *Constr Build Mater.* **225** 132-145
[26] Usman K R, Hainin M R, Satar M K I M, Warid M N M, Usman A, Al-Saffar Z H and Bilema M A 2021*Case Stud. Constr. Mater.* **14** e00474
[27] STAS 2634-1980 (1980) Mortare obişnuite pentru zidărie şi tencuieli. Metode de încercare, *Romanian Standard Association, Bucharest, Romania*
[28] EN 12390-7 (2019) Testing hardened concrete - Part 7: Density of hardened concrete, *Romanian Standard Association, Bucharest, Romania*
[29] EN 1015-10 (2002) Methods of test for mortar for masonry - Part 10: Determination of dry bulk density of hardened mortar, *Romanian Standard Association, Bucharest, Romania*
[30] EN 196-1: (2016) Methods of testing cement. Determination of strength, *Romanian Standard Association, Bucharest, Romania*
[31] EN 1015-11 (2002) Methods of test for mortar for masonry. Determination of flexural and compressive strength of hardened mortar, Romanian Standards Association, Bucharest, Romania (2007)

Olive vegetation waters (OVWs): characteristics, treatments and environmental problems

P De Luca[1], V Sicilia[2], S Candamano[1] and A Macario[2]

[1]Dipartimento di Ingegneria Meccanica, Energetica e Gestionale, University of Calabria, I-87036 Arcavacata di Rende, Italy.
[2]Dipartimento di Ingegneria per l'Ambiente, University of Calabria, Arcavacata di Rende,I-87036 Arcavacata di Rende, Italy.

E-mail: pierantonio.deluca@unical.it

Abstract. The present work aims to analyze the various aspects relating to olive-vegetative waters (OVW) with particular reference to environmental problems. As is known, olive-vegetation waters are one of the most polluting by-products of the oil industry. They are produced in large quantities and have a polluting potential. For this reason, an in-depth and detailed analysis of the production chain is reported up to the correct and lawful disposal of the olive-vegetation waters. The chemical characteristics of vegetation waters are described, with particular reference to polyphenols, which are the most important compounds present in vegetation waters. The treatments, enhancement, perspectives and possible uses in agriculture are reported.

1. Introduction

In recent decades, the most current and debated topic is undoubtedly that of environmental problems [1-3]. In general, the various industrial sectors are certainly increasingly involved in both energy saving and environmental impact issues, especially with regard to emissions and wastewater treatment [4-9]. In particular, food companies have undergone major changes that have revolutionized both the way they produce and treat their waste. There are several techniques that make it possible to reduce or recycle waste. In recent years, many food companies have been dedicating themselves to the development of valorisation strategies, in particular of vegetable waste [10-13].

These are converted into derivatives with high added value, such as ingredients and functional foods, rich in bioactive compounds (for example, fibers, proteins, antioxidants and pigments), but not only [14]. Precisely in the agricultural sector, food waste is often used for the production of fertilizers, through composting or, in the production and industrial processing of feed.

Pomace and vegetation waters (OVWs) represent waste deriving from the production of olive oil [15]. The virgin pomace has a more or less solid consistency and derives from the pulp of olives and is often used to obtain used pomace oil and pomace. The spent pomace can be used as a fuel. OVWs are a more aqueous fraction of waste and represent the main by-product of olive processing. They derive from the constitution water of the drupe and from the water used for the dilution of the olive paste and for washing the machines. Millions of tons of OVWs are produced every year during the olive pressing season. For this reason, the management of OVWs is quite problematic due to the high organic content, the marked acidity and the significant presence of polyphenolic substances which are natural substances with a high antimicrobial activity. The extraction technologies used influence the characteristics of the OVWs. All this means that massive and unregulated disposal can lead to serious problems for the environment. On the other hand, being rich in organic and inorganic products, if properly treated they can be considered a source of products of potential economic interest [16-21]. This work aims to provide a synthesis of the most relevant aspects relating to vegetation waters (OVWs) with particular reference to production processes, characteristics, treatments and uses.

2. By-products in the different olive oil extraction processes

The chemical and physical characteristics of the OVWs and pomace vegetation water depend greatly on the types of process used for the production of oil.

In particular, there are three fundamental processes: traditional; two-phase centrifugation system and three-phase centrifugation system (Figure 1).

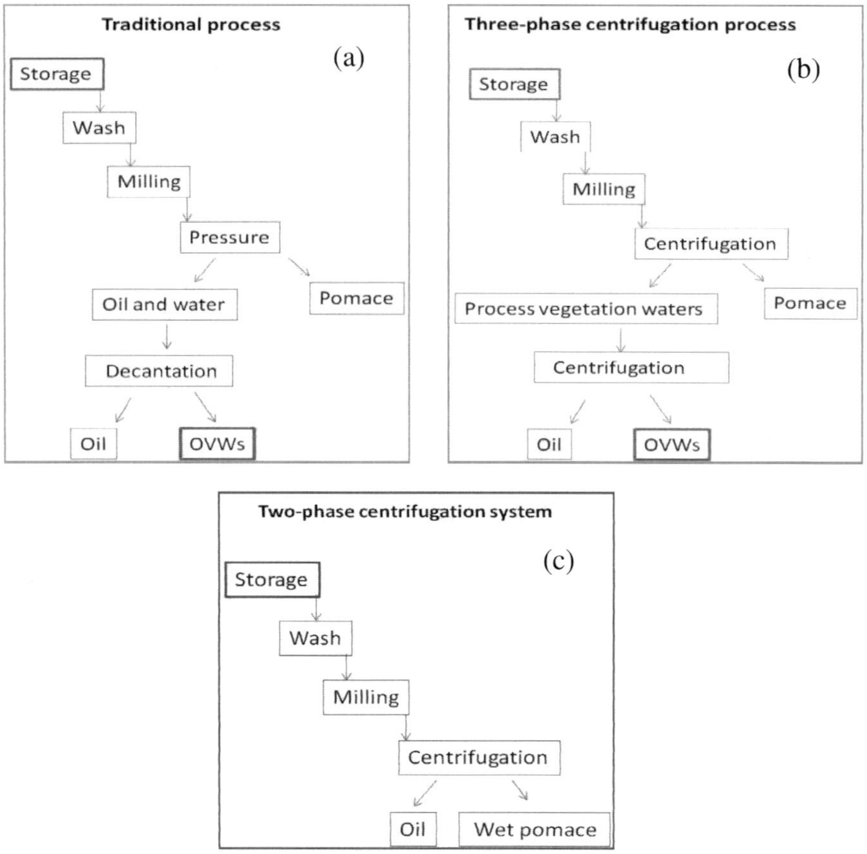

Figure 1. Olive oil production systems.

The first stages, storage, washing and milling or pressing are common to all processes. During the storage phase, the olives are kept cool and well ventilated, sheltered from light and heat sources. The washing is carried out to avoid the accumulation of vegetable waste and to remove any foreign bodies such as residues of earth, sand, etc. The washing operation can be harmful if carried out on olives in an advanced state of ripeness as it can cause an easy decay of the olives. The milling or pressing has the purpose of obtaining a homogeneous paste whose consistency substantially depends on the degree of humidity and is carried out through crushers with breaking of the olives.

Subsequently, the solid-liquid separation can be achieved through:

- Pressing system (Figure 1a) which is the more traditional one, in which the separation of the two phases, solid and liquid, takes place by means of vertical presses which, thanks to the considerable pressure applied and the use of special discs, favor the leakage of the oily must. The further processing of the oily must has the purpose of allowing the separation of the vegetation waters from the oil (OVWs). Traditional milling processes require a variable quantity of water between 40-120 liters per quintal of pressed olives, to be added during the malaxing phase, generating a considerable amount of wastewater.
- In two-phase separation systems (Figure 1c), the pulp is initially separated only into oil and wet pomace. The effluent obtained has a high oil content and a high humidity rate (about 60%). With this process, the wastewater is not very suitable for disposal on agricultural land due to the excessive content of fatty substances [22].
- The three-phase centrifugal system (Figure 1b) involves the use of a horizontal centrifuge, which allows the separation of the oily must from the pomace.

Subsequently the oily must coming out is automatically sent to another centrifugal separator. This extraction system allows the initial separation of the pomace from the olive paste and, subsequently, the removal of the vegetation water from the oil. Working in three phases, pomace with acceptable humidity (48-54%) and high quantities of vegetation water are obtained [22,24].

3. Characteristics of vegetation waters (OVWs)

The vegetation waters have a dark color that can go up to black (Figure 2) and are characterized by a rather intense odor. Basically they consist of an aqueous solution containing organic substances, such as reducing sugars, organic acids, polyalcohols, etc., and minerals such as potassium, phosphorus, calcium. They may also contain suspended solid plant material.

Figure 2. Vegetable waters of olives (OVWs).

The chemical-physical characteristics of the vegetation water depend on the type of process that is used for the extraction of the oil. In particular, the vegetation waters coming from traditional pressure extraction processes have a higher solid component than those coming from more innovative processes such as extraction with centrifuge [25]. The following Table 1 compares the parameters of the vegetation waters of the olives coming from two different extraction processes.

The OVWs have an acidic pH, which varies in a range between 4.5 and 5.9. This acidity is due to the presence of organic acids such as malic and citric acid. The pH variability mainly depends on the storage and ripening period of the olives but is little influenced by the type of extraction process.

The OVWs obtained from an extractive process with centrifugation have, compared to that by pressing, a greater quantity of oil.

Table 1. Parameters of OVWs according to the extraction system [25].

Parameter	Pressure extraction	Centrifugation extraction
pH	5.27	5.23
Dry extract (g L^{-1})	129.7	61.1
Specific weight	1.049	1.020
Oil (g L^{-1})	2.26	5.78
Reducing sugars (g L^{-1})	35.8	15.9
Total polyphenols (g L^{-1})	6.2	2.7
Ashes (g L^{-1})	20.1	6.4
COD (g O_2 L^{-1})	146	85.7
BOD_5 (g O_2 L^{-1})	90.2	28.7

The content of reducing sugars, mainly about 90% glucose and about 10% fructose, is very variable and also depends on the variety, degree of ripeness and storage of the olives and the type of extraction. The quantity of polyphenols in the OVWs is very influenced by the type of extraction in fact with a centrifugation process their concentration decreases [26-29].

Furthermore, their concentration is still influenced by the ripening of the olives. Potassium is the most present mineral element followed by nitrogen, phosphorus and other microelements with more modest concentrations [30]. The values of COD (Chemical Oxigen Demand) and BOD_5 (Biological Oxygen Demand), are always very high, they are much higher for vegetation waters deriving from traditional processes. During storage, the OVWs can undergo fermentation phenomena due to the action of aerobic and anaerobic microorganisms with consequent increase in pH, decrease in BOD5 [31]. The microbial population is mainly made up of bacteria such as Cellulosolytics, Pseudomonas, while nitrifying agents are absent. There are fewer yeasts and fungi, many of which Pectinolytics, Saccharomycetes, but also Penicillum glaucum and Aspergillus niger, while Actinomycetes are absent [32].

4. Polyphenols in OVWs

Polyphenols are a large group of organic compounds of plant origin which are considered an important components of the human diet for their antioxidant action [33,34]. Their most important function is to protect plants from the oxidative risk represented by sunlight and other environmental agents. They contribute to the preservation of the organoleptic properties of food derivatives of vegetable origin such as olive oil. These organic compounds are present in important concentrations in the vegetation water OVWs [35].

During the olive oil extraction process only a small amount of polyphenols of the total present in the olives is transferred to the oil, while an important percentage is retained in the OVWs. The quantity of phenols is very variable in relation to the type of olives, their state of ripeness, storage and the degradation that they had to undergo between harvesting and processing. The fruit of the olives is rich in phenolic compounds, but only 2% of the total content of phenols is found in the oil, a quantity that is still sufficient to protect it from oxidation, while the remainder are lost in the OVWs (about 53 %) and in the pomace (about 45%) [36]. The extraction of biologically active compounds from the OVWs, can be useful for producing natural antioxidants which are of interest to many pharmaceutical and food industries as free radicals are involved in various pathologies, such as cancer and cardiovascular diseases, and in oxidation of foods. The phenolic composition in the OVWs is very different from that of olives. Twenty different biophenols have been well identified in the OVWs. Among the biophenols present in OVWs, the most abundant and most interesting from a nutritional point of view is hydroxytyrosol [37-39].

Other biophenols present in OVWs are tyrosol, caffeic acid and vanillic acid [40-42].

In conclusion, the biophenols present in OVWs show interesting biological activities and could therefore represent an inexpensive source of natural antioxidants, but not yet exploited due to the problems associated with their extraction from the matrices.

5. Treatments of OVWs and perspectives

Purification and enhancement represent two possible solutions to the treatment of OVWs. The first, purification, consists in eliminating some components from the wastewater, through the use of sophisticated technologies [43-47].

The second, enhancement, aims to recover the materials produced from waste through the extraction or use of biotechnologies [48-52]. The different chemical compounds recovered can subsequently be used, by usual chemical synthesis for the preparation of new products [53-56]. The purification treatments give the wastewater particular requirements through physical, biological and chemical processes. Through physical treatments such as distillation, membrane filtration, incineration etc. it is possible to obtain a significant reduction in volumes and the separation of dissolved components in the wastewater without the use of chemical reagents.

The distillation method, however, is very expensive and with a high energy expenditure. Solar distillation appears to be a promising low cost technology [57,58]. Membrane separation processes are innovative, as they are quite effective for reducing the polluting load, thus improving the characteristics of the OVWs. The use of membranes in a sequential design is quite promising. The method is based on two processes, initial ultrafiltration (UF) followed by a nanofiltration (NF) treatment [59]. In this case, the future use of innovative nano materials such as carbon nanotubes which have shown great adsorbing and purifying capacity in many other sectors is particularly interesting [60-67]. Incineration represents a theoretically decisive method for solving the problem of OVWs. In fact, ideally the combustion of the wastewater produces carbon dioxide, water vapor, nitrogen, and ashes made up of mineral salts. However, in practice, the method is not always adopted due to the high energy consumption and the high production of ashes and exhaust gases. The biological processes exploit the action of both aerobic and anaerobic microorganisms to obtain in a short time the self-purification of waste water from the organic component. These treatments, in addition to greater cost-effectiveness, have multiple advantages: low energy consumption due to operating conditions (environmental temperature and pressure), low environmental impact and the possibility of treating large volumes of water. The anaerobic treatment consists in the digestion, by anaerobic bacteria, of the organic load dissolved in the OVWs. This process produces low calorific value biogas containing mainly CO_2, CH_4, H_2S. The recovery of energy in the form of methane makes the anaerobic treatment of OVWS a disposal associated with the valorisation of the wastewater [68-70]. Chemical methods involve the use of chemical reactants and are capable of: oxidizing or reducing polluting compounds, breaking down or eliminating the pathogenic component, precipitating undesirable components in the form of sludge, maintaining the pH within the legal limits. A technology that can be considered green and simple capable of reducing the concentration of pollutants, often even selectively, is represented by the use of microporous materials and resins [71-75].

6. Use of OVWs and environmental problems

The spreading of OVWs on agricultural land must be carefully treated due to:

- high organic load;
- low biodegradability due to the presence of polyphenols with antimicrobial activity.

However, OVWs do not possess effective toxicity but cause just some undesirable effects, however temporary, on the functionality of agroecosystems.

The direct distribution on the agricultural land of the vegetation waters therefore represents the preferential way to proceed with their recovery, and this both for economic reasons and for agro-ecological needs. For this reason the OVWs are regulated by specific legislation [76-79]. The controlled spreading of OVWs leads to an increase in the wet fraction and an enrichment in nutritional elements without showing significant depressive effects on the productivity of the land. The spreading of OVWs on agricultural land has shown that, for dosages between 50-80 m^3 ha^{-1} as the maximum value, no cytotoxic phenomena are found [80-82]. The spreading of vegetation waters in tree crops does not show particular phenomena of phytotoxicity but problems especially as regards the practicability of the land, especially in areas characterized by particular slopes as it is possible to trigger erosion phenomena.

In herbaceous crops, effects on germination have been found if the spreading of vegetation waters occurs with a period of less than 40 days before sowing, therefore the spreading of OVWs on herbaceous crops is desirable in the growth phase.

Due to the high content of nutritional elements such as potassium and phosphorus in OVWs, their use as fertilizers is possible, but their use must be carried out with adequate quantities and according to legislative limits.

The most significant limit of the spreading of OVWS on soils is mainly due to their acidic character with a pH value around 5.

Furthermore, the polyphenols present in OVWs can slow down the transformation and biodegradation processes due to their antimicrobial properties. These effects are logically dependent on the quantities of OVWs that are poured onto the land, although they tend to subside within a few months.

The presence of polyphenols in the soil is not due to a direct toxic action towards plants but to antioxidant and bacteriostatic actions, which can influence the cycles of organic and mineral nutrients present in the soil. In this case, caution in determining the doses and distribution period of OVWS can overcome these drawbacks. Therefore, the OVWs are substantially free of any dangerous substance (pathogens, heavy metals, etc.) and the problem of their use almost exclusively concerns some of the organic components characterized by a marked antimicrobial action and/or a low biodegradability (polyphenols). Contamination of the aquifers is limited to particularly porous soils and only in the event of a very rainy season.

The use of OVWS in agriculture must be carried out in a rational way, paying particular attention to the quantities, times and site intended for spreading which must have the following characteristics:

- water table greater than 10 meters deep;
- respectful distance from inhabited centers and drinking water collection areas;
- exclusion of land with vegetable crops in place, frozen, snow-covered or saturated with water.

The problems related to the shedding of OVWs are limited and can be easily solved, provided that some simple behavioral rules relating to the doses and the time of distribution are observed [83-87]. The characterization is generally carried out through common analytical techniques such as mass spectroscopy, chromatography, FTIR etc. [88-95].

7. Conclusions

OVWs can therefore represent a resource from which fertilizers can be obtained, if used wisely and according to regulatory limits. A more sophisticated way is to recover biophenols from these wastes which are particularly valuable for human health. In the latter case, it is necessary to encourage research towards the study of innovative techniques and processes for more selective and effective recovery of biophenols.

References

[1] Mendoza D L, Benney T M and Boll S 2021 *Sci. Total Environ.* **776** 145778
[2] Mazza S, Aiello D, Macario A and De Luca P 2020 *Environments*, 7 **5** Article number 37
[3] Markert B A, Breure A M and Zechmeister H G (Eds.) Bioindicators & Biomonitors: Principles, Concepts, and Applications; Trace metals and other contaminants in the environment; Elsevier: Amsterdam, The Netherlands; Boston, MA, USA, 2003; ISBN 978-0-08-044177-1
[4] Siatou A, Manali A and Gikas P 2020 *Water* **12** 1204

[5] De Luca P, Foglia P, Siciliano C, B.Nagy J, Macario A. 2019 *Environments* 6 **101** doi: 10.3390/ environments 6090101
[6] Paranychianakis N V, Salgot M, Snyder S A, Angelakis A N. 2015 *Crit. Rev. Environ. Sci. Technol.* **45** 1409–1468
[7] Aboltins A, Karps O, Palabinskis J. 2017 *Int. Sci. Conf. Rural Dev.*, 193–198
[8] Spinosa L. Wastewater Sludge: A Global Overview of the Current Status and Future Prospects, 2nd ed.; IWA Publishing: London, UK, 2011
[9] Jaramillo M F, Restrepo I. 2017 *Sustainability* **9** 1734
[10] Minuti L, Ballerini E, Barattucci A, Bonaccorsi P M, Di Gioia M L, Leggio A, Siciliano C, Temperini A. 2015 *Tetrahedron* **71** 3253-3262
[11] Alam M W, Pandey P, Khan F, Souayeh B, Farhan M. 2020. *Int. J. Environ. Res. Public Health* **17** 7468
[12] Aiello D, Siciliano C, Mazzotti F, Di Donna L, Athanassopoulos C M, Napoli A. 2018 *RSC Adv.* **8** 36104-3611
[13] Muhamad Ng SN, Idrus S, Ahsan A, Tuan Mohd Marzuki T N, Mahat SB. 2021 *Membranes* **11** 456
[14] Aiello D, Siciliano C, Mazzotti F, Di Donna L, Risoluti R, Napoli A. 2020 *Molecules* **25** 1485
[15] Romero-García J M, Niño L., Martínez-Patiño C, Álvarez C, Castro E, Negro, M J. 2014 *Bioresour. Technol.* **159** 421–432
[16] Del Mar Contreras M, Romero I, Moya M, Castro E. 2020 *Process Biochem.* **97** 43–56
[17] Roig, A, Cayuela M L, Sánchez-Monedero M A. 2006 *Waste Manag.* **26** 960–969
[18] Şahin, S, Bilgin M. 2018 *J. Sci. Food Agric.* **98** 1271–1279
[19] Visioli F, Romani A, Mulinacci N, Zarini S, Conte D, Vincieri F F. et al., 1999. *Journal of Agricultural and Food Chemistry* **47** 3397-3401
[20] Siciliano C, De Marco R, Guidi L E, Spinella M, Liguori A. 2012 *J. Org. Chem.* **77** 10575–10582.
[21] De Marco R, Di Gioia M L, Leggio A, Liguori A, Perri F, Siciliano C, Viscomi M C. 2010 *Amino Acids* **38** 691–700
[22] Alburquerque J A, Gonzálvez J, García D, Cegarra J. 2004 *Bioresour. Technol.* **91** 195–200
[23] Tamborrino A, Leone A, Romaniello R, Catalano P, Bianchi B. 2015 *Biosyst. Eng.* **129** 160–168
[24] Squeo G, Tamborrino A, Pasqualone A, Leone A, Paradiso V M, Summo C, Caponio F. 2017 *Food Bioprocess Technol.* **10** 592–602
[25] Di Giovacchino L, Mascolo A, Seghetti L, 1988. *La Rivista delle Sostanze Grasse.* Vol. LXV
[26] Lozano-Sánchez J, Giambanelli E, Quirantes-Piné R, Cerretani L, Bendini A, Segura-Carretero A, Fernández-Gutiérrez, A. 2011 *J. Agric. Food Chem.* **59** 11491–11500
[27] De Ceglie C, Abbattista R, Losito I, Castellaneta A, Calvano C D, Bianco G, Palmisano F, Cataldi T R I. 2020 *J. Agric. Food Chem.* **68** 3171–3183
[28] .Menchetti L, Taticchi A, Esposto S, Servili M, Ranucci D, Branciari R, Miraglia D. 2020 *LWT Food Sci. Technol.* 129 **2** 109648109648. DOI:10.1016/j.lwt.2020.109648
[29] Fasolato L, Cardazzo B, Balzan S, Carraro L, Taticchi A, Montemurro F, Novelli E. 2015 *Ital. J. Food Saf.* **4** 75–77
[30] Balice V, Carrieri C, Cera O. 1990. *Riv. Ital. Sost. Grasse* **67** 9-1
[31] Evci E, Yılmaz Ö, Bekfelavi E Y, Kuş N S. 2019 *SN Appl. Sci.* **1** 1206
[32] Miraglia D, Castrica M, Menchetti L, Esposto S, Branciari R, Ranucci D, Urbani S, Sordini B, Veneziani G, Servili M. 2020 *Foods* **9** 1647
[33] Ventura G, Calvano C D, Abbattista R, Bianco M, De Ceglie C, Losito I, Palmisano F, Cataldi T R I. 2019 *Rapid Commun. Mass Spectrom.* **33** 1670–1681
[34] Nocella C, Cammisotto V, Fianchini L, D'Amico A, Novo M, Castellani V, Stefanini L, Violi F, Carnevale R. 2018 *Endocr. Metab. Immune Disord. Drug Targets* **18** 4–13

[35] Ruzzolini J, Peppicelli S, Andreucci E, Bianchini F, Scardigli A, Romani A, La Marca G, Nediani C, Calorini L. 2018 *Nutrients* **10** 1950
[36] Ryan D, Robards K. 1998 *Analyst* **123** 31R-44R
[37] Rodis P S, Karathanos V T, Mantzavinou A. 2002. *Journal of Agricultural and Food Chemistry* **50** 596-601
[38] Visioli F, Romani A, Mulinacci N, Zarini S, Conte D, Vincieri F F. et al., 1999. *Journal of Agricultural and Food Chemistry* **47** 3397-3401
[39] Aruoma O I, Deiane M, Jenner A, Halliwell B, Kaur M, Bnni S, et al. 1998. *Journal of Agricultural and Food Chemistry* **46** 5181- 5187
[40] Bisignano G, Tomaino A, Lo Cascio R, Crisafi G, Uccella N, Saija A. 1999. *Journal of Pharmacy and Pharmacology* **51** 971-974
[41] De La Puerta R, Gutierrez V R, Hoult J R S. 1999. *Biochemical Pharmacology* **57** 445-449
[42] Liu M J, Li J X, Guo H Z, Lee K M, Qin L, Chan K M. 2003. *Life Science* **73** 883-892
[43] Pietta P G. 2000 *Journal of Natural Products* **63** 1035-1042
[44] Paraskeva P, Diamadopoulos E. 2006 *J. Chem. Technol. Biotechnol.* **81** 1475–1485
[45] Morillo J A, Antizar-Ladislao B, Monteoliva-Sánchez M, Ramos-Cormenzana A, Russell N J. 2009 *Appl. Microbiol. Biotechnol.* **82** 25–39
[46] Servili M, Esposto S, Veneziani G, Urbani S, Taticchi A, Di Maio I, Selvaggini R, Sordini B, Montedoro G F. 2011 *Food Chem.* **124** 1308–1315
[47] Obied H K, Allen M S, Bedgood D R, Prenzler P D, Robards K. 2005 *J. Agric. Food Chem.* **53** 9911–9920
[48] Bhatnagar A, Kaczala F, Hogland W, Marques M, Paraskeva C A, Papadakis V G, Sillanpää M. 2014 *Environ. Sci. Pollut. Res.* **21** 268–298
[49] Dermeche S, Nadour M, Larroche C, Moulti-Mati F, Michaud P. 2013 *Process. Biochem.* **48** 1532–1552
[50] Rahmanian N, Jafari S M, Galanakis C M. 2014 *J. Am. Oil Chem. Soc.* **91** 1–18
[51] Alcaide E M, Nefzaoui A. 1996 *Int. Biodeterior. Biodegrad.* **38** 227–235
[52] Zbakh H, El Abbassi A. 2012 *J. Funct. Foods* **4** 53–65
[53] Antoniou A I, Pepe D A, Aiello D, Siciliano C, Athanassopoulos C M. 2016 *J. Org. Chem.* **81** 4353–4358
[54] Temperini A, Aiello D, Mazzotti F, Athanassopoulos C M, De Luca P, Siciliano C. 2020 *Molecules* **25** 1313
[55] De Marco R, Di Gioia M L, Liguori A, Perri F, Siciliano C, Spinella M. 2011 *Tetrahedron* ***67*** 9708–9714
[56] Minuti L, Barattucci A, Bonaccorsi P M, Di Gioia M L, Leggio A., Siciliano C, Temperini A. 2013 *Org. Lett.* ***15*** 3906–3909.
[57] Khdair A, Abu-Rumman G. 2020 *Processes* **8** 671
[58] Ena A, Pintucci C, Carlozzi P. 2012 *Journal of Biotechnology* **157** 573–577
[59] Cassano A, Conidi C, Giorno L, Drioli E. 2013 *Journal of Hazardous Materials* 248-249C **1** 185–193
[60] Upadhyayula V K K, Deng S, Mitchell M C, Smith G B. 2009 *Sci. Total Environ.* **408** 1–13
[61] De Luca P, Siciliano C, Macario A, B.Nagy J. 202*1* *Materials* 14 **9** 2118; https://doi.org/10.3390/ma14092118
[62] Ma L, Dong X, Chen M, Zhu L, Wang C, Yang F, Dong Y. 2017 *Membranes* **7** 16
[63] De Luca P, Chiodo A, Macario A, Siciliano C, B.Nagy J. 2021 *Appl. Sci.* 11 **4** 1687 https://doi.org/10.3390/app11041687
[64] Mauter M S, Elimelech M.. 2008 *Environ. Sci. Technol.* **42** 5843–5859
[65] Sarkar B, Mandal S, Tsang Y F, Kumar P, Kim K H, Ok Y S. 2018 *Sci. Total Environ.* **612** 561–581
[66] De Luca P, B.Nagy J. 2020 Materials 13 **23** 5508; https://doi.org/10.3390/ma13235508

[67] De Benedetto C, Macario A, Siciliano C, B.Nagy J, De Luca P. 2020 *Materials* 13 **12** 2757; https://doi.org/10.3390/ma13122757
[68] Bertin L, Berselli S, Fava F, Petrangeli-Papini M, Marchetti L. 2004 *Water research* **38** 3167–3178
[69] Bertin L, Colao M C , Ruzzi M, Fava F. 2004 *FEMS microbiology ecology* **48** 413-423
[70] Sayadi S, Allouche N, Jaoua M, Aloui F., 2000 *Process Biochemistry* **35** 725–735
[71] .De Luca P, Poulsen T G, Salituro A, Tedeschi A, Vuono D, Kònya Z, Madaràsz D, B.Nagy, J. 2015 *J. Therm. Anal. Calorim.* **122** 1257–1267
[72] De Luca P, Mastroianni C, Siciliano C, B.Nagy J, Macario A. 2019 *Gels* **5** 42; doi:10.3390/gels5030042
[73] Kuznicki S M, Bell V A, Nair S, Hillhouse H W, Jacubinas R M, Braunbarth C M, Toby B H, Tsapatsis M. 2001 *Nature* **412** 720-724
[74] Wang S, Peng Y. 2010 *Chem. Eng. J.* **156** 11–24
[75] De Luca P, Vuono D, Filice M. 2009 *Environmental Engineering and Management Journal.* 8 **5** 1009-1015
[76] PROSODOL and Life 2012*Integrated Strategy of Actions, Measures and Means Suitable for Mediterranean Countries* Analysis of national and European legislative frameworks for Olive Oil Waste and Soil Protection. Available at:http://www.prosodol.gr/sites/prosodol.gr/files/ACT15_1.pdf
[77] Inglezakis V J, Moreno J L, Doula M K. 2012 *Int. J. Chem. Environ. Eng. Syst.* **3** 65–77
[78] European Environment Agency. 2009 *EEA Rep.* 1–68
[79] Cini E, Regis F. 2000. Smaltimento-recupero delle acque di vegetazione dei frantoi oleari. Da “Smaltimento e riutilizzo dei reflui dai frantoi”. Ed.ARSIA-Regione Toscana
[80] Catalano M. 1989. International seminar on “Trattamento Acque reflue oleifici. Lecce, 16-17 novembre1989
[81] Cicolani B., Seghetti L., D’Alfonso S., Di Giovacchino L., 1993. *L’Informatore Agrario* **34** 69–75
[82] Bonari E., 1990. Primi risultati sperimentali sullo spargimento di acque di vegetazione sul terreno agrario in Toscana. A International seminar on “Olio di oliva e olive da tavola: tecnologia e qualità”: Città Sant'Angelo, Pescara, 25-28 aprile 1990
[83] Bonari E, Ceccarini L, 1991. *L’informatore agrario* **13** 49–57
[84] Justino C I L, Pereira R, Freitas A C, Rocha-Santos T A P, Panteleitchouk T S L, Duarte A.C 2012 *Ecotoxicology* **21** 615–629
[85] Ouzounidou G, Zervakis G, Gaitis F. 2010 *Terr. Aquat. Environ. Toxicol.* **4** 21–38
[86] Torrecilla J S. 2010 In Olives and Olive Oil in Health and Disease Prevention; Elsevier: Amsterdam, The Netherlands 357–365. ISBN 9780123744203
[87] Barbera A C, Maucieri C, Cavallaro V, Ioppolo A, Spagna G. 2013 *Agric. Water Manag.* **119** 43–53
[88] Abbattista R, Ventura G, Calvano C D, Cataldi T R I, Losito I.. 2021 *Foods* **10** 1236
[89] Aiello D, Giambona A, Leto F, Passarello C, Damiani G, Maggio A, Siciliano C, Napoli A. 2018 *Sci. Rep.* **8** Article number 10973
[90] Mattonai M, Vinci A, Degano I, Ribechini E, Franceschi M, Modugno F. 2019 *Nat. Prod. Res.* **33** 3171–3175
[91] Hafidi M, Amir S, Revel J.C. 2005 *Process. Biochem.* **40** 2615–2622
[92] Tsimidou M, Papadopoulos G, Boskou D. 1992 *Food Chemistry* **44** 53-60
[93] Burduhos Nergiş D P, Nejneru C, Achiţei D C, Cimpoieşu N and Bejinariu C 2018 *IOP Conf. Ser.: Mater. Sci. Eng.* **374** 012040
[94] Nejneru C, Perju MC, Sandu AV, Axinte M, Quaranta M, Sandu I, Costea M and Abdullah MMAB 2016 *Revista de chimie* **67(6)** 506-511.
[95] Vizureanu P and Agop M 2007 *Materials Transactions* **48(11)** 3021-3023. DOI: 10.2320/matertrans.MRP2007118.

Millisecond pulsed laser welding of AISI 316 stainless steel

E M Stanciu[1], A Pascu[1,*], I C Roată[1], C Iatan[1], E R Moldovan[1] and M H Tierean[1]

[1]Transilvania University of Brasov, Department of Materials Engineering and Welding, 1 Universității, Brasov, ROMANIA

Email: alexandru.pascu@unitbv.ro

Abstract. Nowadays, there is an increasing demand for AISI 316 stainless steel for automotive, medical, aerospace, for tool manufacturing and for nuclear industry. Compared with continuous beam laser welding, pulse temporal shape, pulse duration and the repetition rate are fundamental parameters that have a direct influence on the quality on the welded joint [2]. The laser welding of AISI 316 stainless steel thin sheets by millisecond pulsed laser will be analysed within this research by addressing to pulsed laser parameters influence on the geometry and appearing of the welding joint. The influence of laser power, pulse duration and repetition rate were determined by tracking fusion lines on specimen's plates and butt weld stainless steel welding were obtained by using the determined optimal welding parameters. All the parameters have a major influence on the geometrical appearance of the weld bead.

1. Introduction

In the last decades, the word laser has become synonym with precision, speed and high end technology. Unlike classic joining methods which entail well known disadvantages such as high stress and strain due to consistent heat input, dependable welding quality according with personnel skill, reduced efficiency and limited usability in thin plates, laser welding appeared to grow into 21st century solution for joining approach.

Adopted in various industrial areas from automotive and marine sectors to aerospace, pharmaceutical, electronics and offshore, where corrosion resistance is a vital material performance, stainless steel encounters an increase in demand. Thus, austenitic stainless steel such as AISI 316 has become subject for plenty research papers where material behaviour through laser welding is studied in order to determine optimal technological parameters.

It is important to mention that laser joining technique can be performed in two distinct wave modes essentially continuous for deep welding joints and pulsed for thin components where thorough oversight of the energy is needed. The mechanism behind a pulsed welding seam described by low residual stress, solid mechanical properties and proper appearance, consists in a balanced overlapping sequence of high crests of energy, sharply characterized by pulse energy and density [1-3].

From the perspective of weaknesses, pulsed welding is susceptible to thermal cracking, due to the fast cooling. Related to this, Zhang et all [1] proposed to model the pulse shape aiming to control crystallization process on cooling. By virtue of the extensive thermal expansion coefficient and poor thermal conductivity of 300 series of stainless steel, laser welded structures are exposed to residual stress and strain. In this context, Yan et all [4] used Goldak`s double-ellipsoidal in FE method to describe how the amplitude and frequency of the oscillation laser beam can influence residual stress in 316 SS. Although stainless steel is mainly used for its corrosion resistance, studies over austenitic stainless steels have shown that corrosion resistance performance in the HAZ is influenced by laser heat input.[5] Worth of mentioning is the influence of protection gas over the hardness of the LBW specimen. Chatterjee et all. [6] indicate a lower hardness value when using argon as shielding gas. Furthermore, it has been reported that for LBW seam, the width of the joint is only influenced by the laser power, while pulse width has the biggest impact over the ultimate strength of 316 SS [3]. Up to date, there are numerous reports related with stainless steel welding but the process parameters influence for pulsed beam have not been sufficiently studied [7-11]. The overall welding quality depend of a multitude of factors and variation of process parameters.

In this paper the influence of the main pulsed laser welding process parameters will be determined as function of the weld bead geometry and overall mechanical resistance.

2. Materials and methods

2.1. Materials and equipment's

Stainless steel AISI 316L was used as base material for the experimental fusion lines and butt welding. Plain sheets of 100x100x1 mm were used as coupon for the laser processing. The chemical composition of the material is presented in Table 1.

Table 1. Chemical composition of the base material.

Material	Element wt. (%)								
	C %	Si %	Mn %	P %	S %	Cr %	Ni %	Mo %	N %
AISI 316L	0.03	0.75	2	0.0045	0.003	16-18	10-14	2-3	0.10

All the experiments we carried out using pulsed laser, Trumpf TruPulse 556, with a peak pulse power of 10 kW and a wavelength of 1064 nm. The optical system consists in an Precitec YW50 coaxial welding head with 150 mm focal length being manipulated by a CLOOS 7 axes welding robot. The sampes were positioned and clamped using a special designed table as depicted in Figure 1. The Argon gas shielding was provided from the top and at the root of the weld bead in order to prevent any external contamination.

Figure 1. Details of the experimental frame.

The experimental tests were conducted using pulsed beam with 0.7 mm spot diameter in focal on the surface of the material. A total of 18 fusion lines and one welding were realised using a constant processing speed of 45 cm / min. The input variables were the laser power, pulse duration and pulse repetition rate using one single factor at a time approach. The process parameters window was chosen based on preliminary tests. Gradual increasing of the laser power, laser pulse duration and laser repetition rate was used for optimisation of the laser welding process as presented in Table 2, 3 and 4.

Table 2. Variation of the laser power.

Parameters / sample	1.1	1.2	1.3	1.4	1.5	1.6
Laser power [w]	900	1000	1100	1200	1300	1400
Pulse duration [ms]	7	7	7	7	7	7
Repetition rate [Hz]	40	40	40	40	40	40
Energy [J]	6.3	7	7.7	8.4	9.1	9.8
Welding speed [cm/min]	45	45	45	45	45	45

Table 3. Variation of the laser pulse duration.

Parameters / sample	2.1	2.2	2.3	2.4	2.5	2.6
Laser power [w]	1300	1300	1300	1300	1300	1300
Pulse duration [ms]	5	5.5	6	6.5	7	7,5
Repetition rate [Hz]	40	40	40	40	40	40
Energy [J]	6.5	7.1	7.8	8.3	9.1	9.8
Welding speed [cm/min]	45	45	45	45	45	45

Table 4. Variation of the laser repetition rate.

Parameters / sample	3.1	3.2	3.3	3.4	3.5	3.6
Laser power [w]	1300	1300	1300	1300	1300	1300
Pulse duration [ms]	7	7	7	7	7	7
Repetition rate [Hz]	15	20	25	30	35	40
Energy [J]	9.1	9.1	9.1	9.1	9.1	9.1
Welding speed [cm/min]	45	45	45	45	45	45

2.2. Methods

Samples for the metallographic investigation were cut and prepared using standardised methods (polishing with sand paper successively from 180 to 2000 grit and polished on velvet cloth with 0.3 μm Alumina suspension). Further, the samples were electrochemically etched using 10% solution of oxalic acid for 20 seconds at 5V and 500 mA.

The macro images of the samples were captured using the LEICA EZ4 stereo microscope and the microstructures of the fusion lines and welding joints were analysed by optical microscopy using the LEICA DML inverted microscope. The mechanical behaviour of the welded specimen was determined using a universal testing machine WDW-150 S (150 kN).

3. Results and discussion

All the samples were cross-section analysed for determining the best welding geometry. As in case of any welding, the aim is to obtain a full penetration and a minimal width of the melting area by using the smalless amount of energy. The energy is responsible for the dimension of the heat affected zone of any laser processing. The information about the laser power and the power density for obtaining a correct weld bead profile can be observed in Figure 2. The macrographs from Figure 2, samples 1.1 to 1.6 reveal that melt depth increase at higher laser power at an almost exponential rate. It is an expected result and it can clearly see that a full penetration is obtained only at 1400 W.

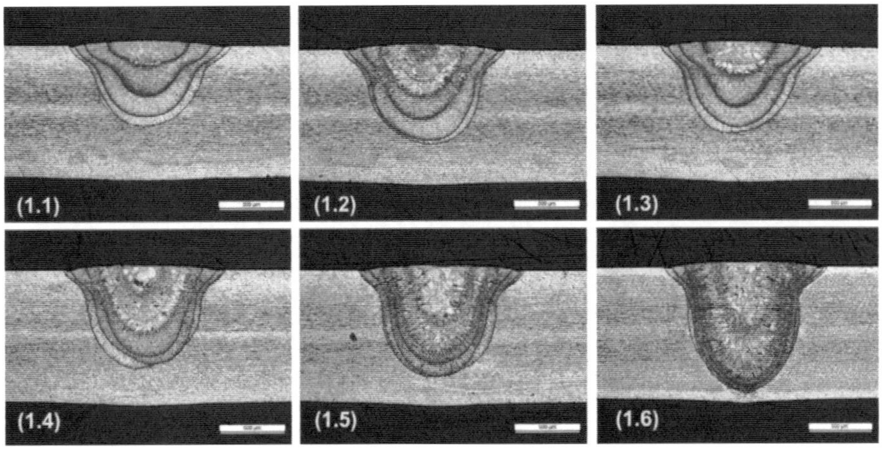

Figure 2. Laser power influence on the melt geometry. Cross section of samples 1.1 to 1.6.

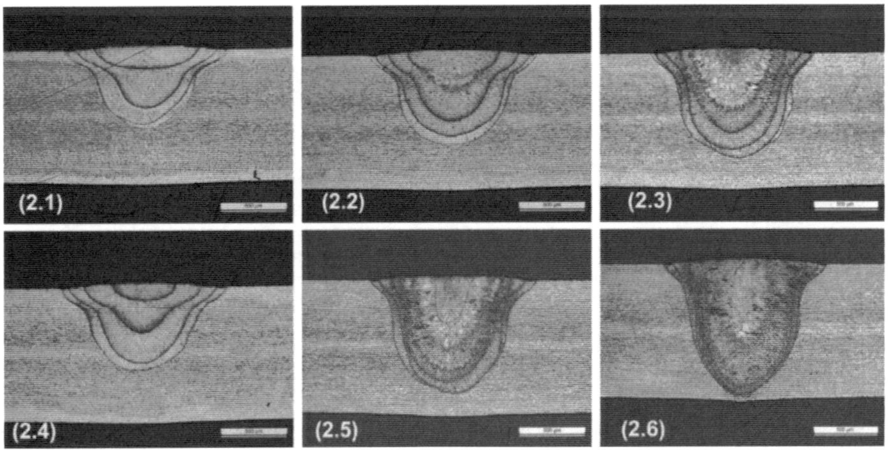

Figure 3. Pulse duration influence on the melt geometry. Cross section of samples 2.1 to 2.6.

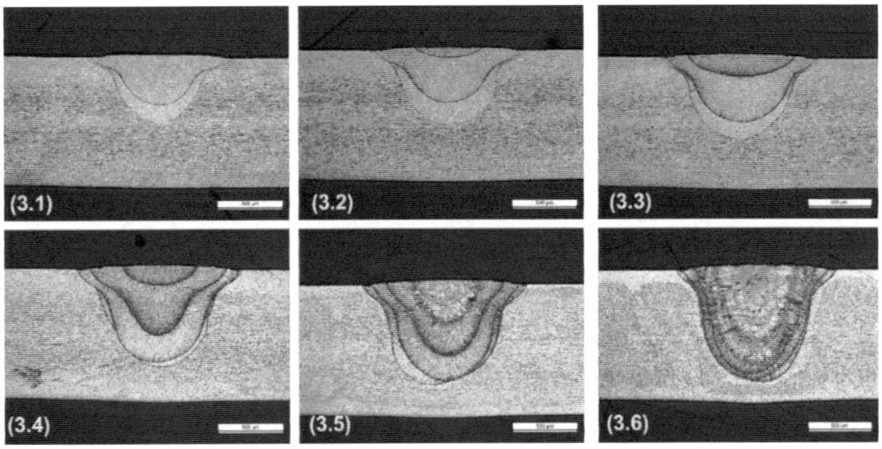

Figure 4. Laser repetition rate influence on the melt geometry. Cross section of samples 3.1 to 3.6.

Similar results were also obtained for pulse duration and repetition rate. Figure 3 shows that an almost similar melt geometry can be obtained by increasing the pulse duration. Increasing of pulse duration leads to increasing of power density that will increase the penetration depth and melt width. The pulse energy [J] is one of the most important parameters in laser processing (pulse energy is the average power * pulse duration). As can be observed in Table 2 and 3, samples 1.6 and 2.6 were realised with the same laser energy but with various power and pulse duration. The results show that at higher pulse duration the melt depth will increase even if the overall pulse energy is the same (Figure 4).

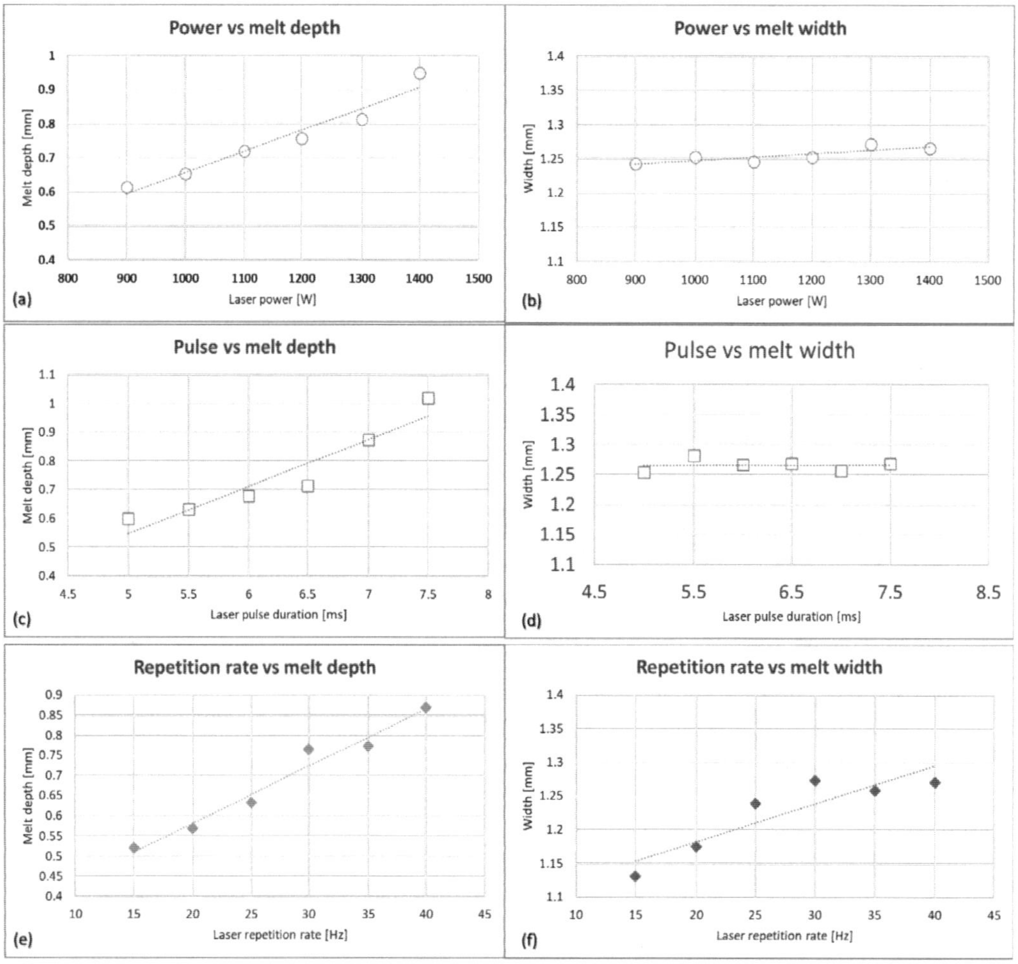

Figure 5. Graphical representation of the process parameters influence on the melt depth and width, and indication of linear trendline (a) Power vs melt depth, (b) laser power vs width, (c) pulse duration vs melt depth, (d) pulse duration vs width, (e) pulse repetition rate vs melt depth, (f) pulse repetition rate vs melt width.

The pulse energy must be corelated with the repetition rate as determined from Figure 5 e and f. The melt depth and width are influenced by the repetition rate. The melt depth and width present low values even if a high pulse energy of 9.1 was used for the samples 3.1 to 3.6. Moreover, the laser repetition rate during the welding process must be related with the welding speed. It is obvious that if we increase the repetition rate, more pulses will interact with the same surface area if the speed is constant. Our experiment reveals that at constant speed we can obtain a full penetration by increasing the repetition rate.

The repetition rate is the parameter that has a major influence on the microstructure of the melted area. A high repetition rate means that total amount of laser energy per surface area will be higher.

Figure 6 shows the microstructure of the sample 3.5. realised with 35 Hz repetition rate. The microstructure of the SS 316L presents a fully austenitic matrix with polyhedral grains. The fast-cooling phenomena during the laser processing and the multiple interaction of the laser pulses with the material leads to formation of multiple solidification patterns (Figure 6 a). Lathy and skeletal ferrite, visible as darker structures in Figure 6b are formed during the melting-solidifying and remelting process at the interaction with laser pulses.

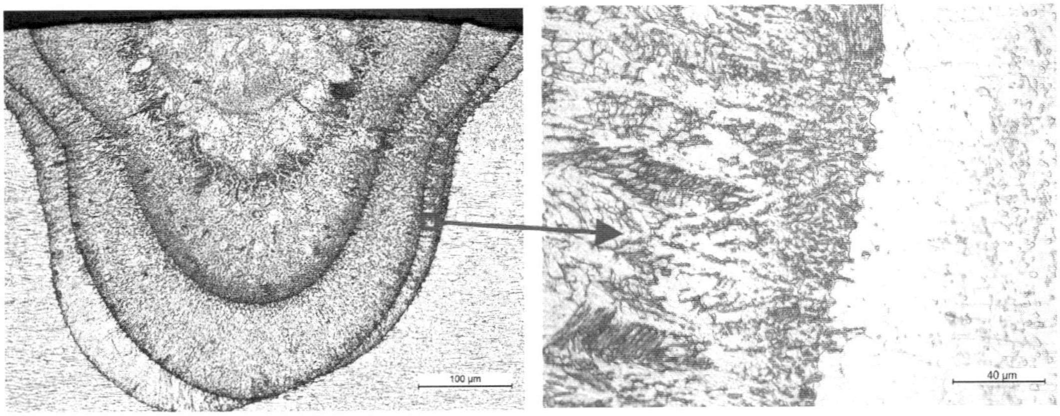

Figure 6. Macro (left) and microstructure (right) of the sample 3.5.

Using the graphical representation from figure 5 was determined that optimal parameters for welding of AISI 316 L stainless steel are: laser power of 1300 W at 7ms pulse duration and 40 Hz repetition rate (Figure 7). Welding speed of 45 cm/min was used for all the experimental tests. The top and the root of the welding was shielded with pure argon gas. The butt weld joint design is simple but no-gap positioning between components is mandatory for thin sheets laser welding. Five welding joints has been realised to validate the process repeatability.

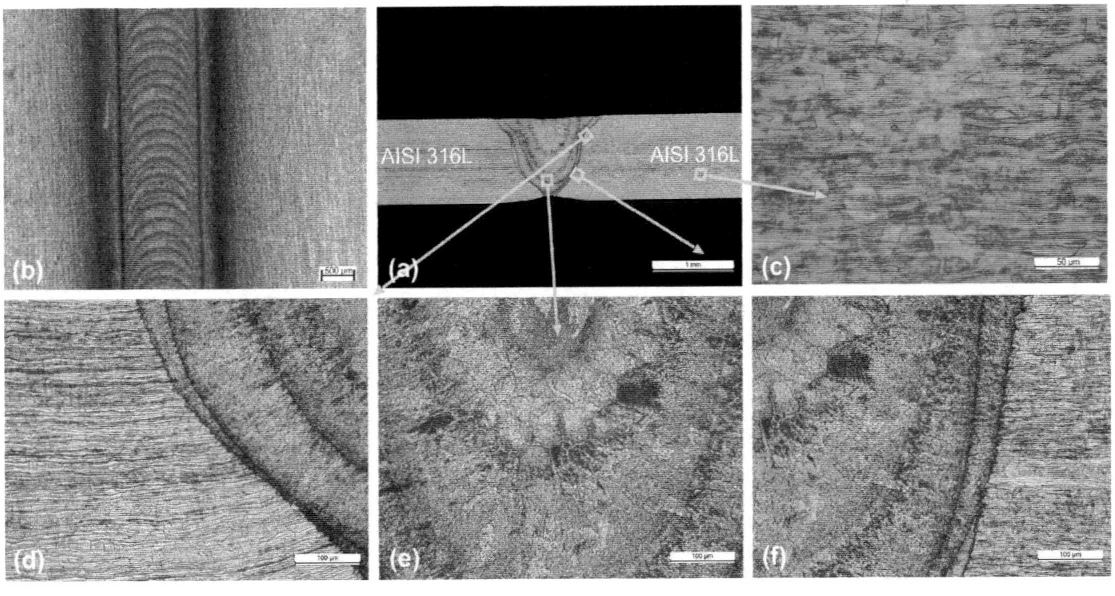

Figure 7. Pulsed laser welding of AISI 316L, a) weld bead profile, b) top view of the weld joint, c) base material, d) and f) detail of the weld bead – base material interface, e) detail of the weld area.

Defect free welded joint was obtained using the determined parameters. Figure 7 shows the overall appearance of the weld joint and details from the interface of melt zone and base material. The weld is characterised by full penetration and no visible heat affected zone.

The tensile behaviour of the welded was performed on specimen of 10 mm wide and 200 mm length. The tensile testing samples were fractured far away from the HAZ or welded joint. The specimen was fractured according with the stress strain curve from Figure 8.

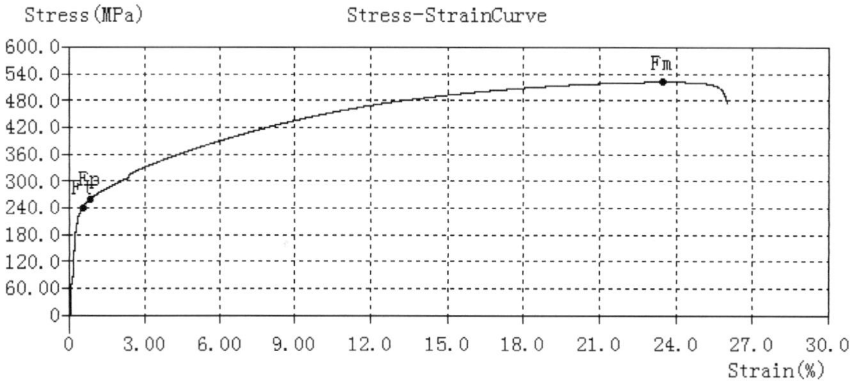

Figure 8. Stress-strain obtained on AISI 316 L laser welding.

4. Conclusion

This study is intended to restrain the interval field of pulsed laser welding, especially for AISI 316L stainless steel. The process parameters play an important role in controlling the geometrical profile of the weld bead.

The 18 tests samples reveal that laser process parameters have a major influence on the geometry and microstructure of the fusion lines. Laser energy of minimum 9.1 Joule is necessary to obtain a full penetration welding. As determined, the minimum energy can be obtained by modifying the laser power or the laser pulse duration, the best parameters for welding of AISI 316L being as following: Laser power (1300W), Pulse duration (7ms), Repetition rate (40Hz).

References

[1] Peilei Z, Zhiyuan J, Zhishui Y, Haichuan S, Shaowei L, Di W, Hua Y, Xin Y, Jieshi C, Fuxin W, and Yingtao T 2021 *Optics & Laser Technology* **140** 107094
[2] Chludzinskia M, Santosa R E, Churiaquea C, Fernandez-Vidalb S R, Ortega M, and Sanchez-Amayaa J M 2021 *Optics and Laser Technology* **134** 106583
[3] Nikhil K, Manidipto M, and Asish B 2017 *Optics & LaserTechnology* **88** 24–39
[4] Shenghong Y, Zheng M, Bo C, Caiwang T, Xiaoguo S, and Guodong W 2022 *Optics & Laser Technology* **145** 107493
[5] Jiahao Z, Ke H, Jiayi Z, Shuyao D, and Xiaohong Z 2022 *Materials Today Communications* **30** 103054
[6] Suman C, Siba S M, Vijay B, Brahma N U, Khushvinder S B, and Joji T 2019 *Optics and Laser Technology* **117** 186–199
[7] Ciubară A, Burlea Ș L, Axinte M, Cimpoeșu R, Chicet DL, Manole V, Burlea G and Cimpoeșu N 2018 *IOP Conference Series: Materials Science and Engineering* **374(1)** 012066.
[8] Abtew MA, Boussu F, Bruniaux P, Loghin C, Cristian I, Chen Y and Wang L 2018 *AIP Conference Proceedings* **1960** 020001. doi.org/10.1063/1.5034802.

[9] Cimpoesu N, Stanciu S, Vizureanu P, Cimpoesu R, Achitei DC and Ionita I 2014 Journal of mining and metallurgy section B-metallurgy **50(1)** 69-76. DOI: 10.2298/JMMB121206010C.
[10] Nejneru C, Cristina Perju M, Doru Burduhos Nergis D, Victor Sandu A and Bejinariu C 2019 *Revista de Chimie* **70** 3597–3602
[11] Nergis, DDB, Vizureanu P, Sandu AV, Nergis, DPB and Bejinariu C 2022 *Materials* **15(1)** 202. DOI: 10.3390/ma15010202.

IOP Conf. Series: Materials Science and Engineering 1251 (2022) 012013 doi:10.1088/1757-899X/1251/1/012013

Recycling plastic wastes in order to obtain new building materials

L Plesa[1], L D Manea[1] and R Istoan[1]

[1]Technical University of Cluj-Napoca, 28 Memorandumului Street, 400114, Cluj-Napoca, Romania

Email: luminitamolnar@ccm.utcluj.ro

Abstract. This paper presents the development of new types of construction materials using plastic waste. According to the European Court, plastic waste from various sectors such as the automotive, electronics, agriculture and construction sectors is a growing problem and is damaging to both terrestrial and marine media. Thus, every year, between 4.8 and 12.7 million tons of plastics reach the oceans. According to most of the research studies, plastic wastes have the potential to replace the natural aggregate at specific percentages. This paper proposes the reuse of this waste as reinforcement in masonry mortars, using plastic waste in various proportions. The experimental program follows the characteristics of masonry mortars both in terms of the behaviour of fresh mortar and in terms of mechanical strength. The results obtained showed a better performance of mortars reinforced with plastic strips, both in terms of flexural strength and compressive strength.

1. Introduction

One of the main objectives of all world governments is to find a solution to pollution by various types of waste. According to the European Commission, around 5 tonnes of waste is produced annually in Europe, of which only 38% is recycled and more than 60% of household waste ends up in landfills in some EU countries [1]. In recent years, waste generation has increased significantly, and waste management is therefore essential for sustainable development [2-4]. Due to the presence of waste, the environment has been affected both in terms of the space it occupies and in terms of air pollution from the release of carbon dioxide [5-8].

Plastic is one of the most widely used materials, but at the same time it can have a negative impact on human health as well as on the world environment. Plastic is one of the most polluting wastes, and there is a global drive to reduce its impact through several directives [9-11]. At European level, according to statistics, around 26 million tons of plastic are generated every year, 80% of which is dumped in water (rivers, lakes, seas, and oceans). To implement the proposed solutions as efficiently as possible, Romania has adopted Directive (EU) 2019/904 on the reduction of the environmental impact of certain plastic products [12]. According to the European Directive, among the single-use plastic products that are considered for recycling or even reuse is the following: fishing gear; beverage bottles and bottle caps; chopsticks, chip packets, wet wipes, sanitary napkins, cutlery, straws, drink stirrers, drinking glasses, food containers [2].

Building materials is an industry where innovation and new product development is constantly evolving. In this regard, there are many studies that attempt to reuse plastic waste in the production of mortars and concrete with plastic waste [13].

New research is on the reusing plastic wastes to develop bricks from PET bottles wastes. Bricks made in this way contain 25% recycled plastic, but scientists are working on increasing the proportion (50%, 75% and 100%) of plastic in a brick [3].

Another approach to reusing PET bottle waste is to fill it with sand, obtaining mechanical strengths comparable to those obtained on bricks.

According to the bibliometric analysis, which was made for the last fifth years, it could be seen that the interest in research of plastic wastes in the building materials is increasing (figure 1). Thus, in 2017, 48 scientific papers were published, reaching 112 scientific papers in 2021, with over 27.614% [4-7].

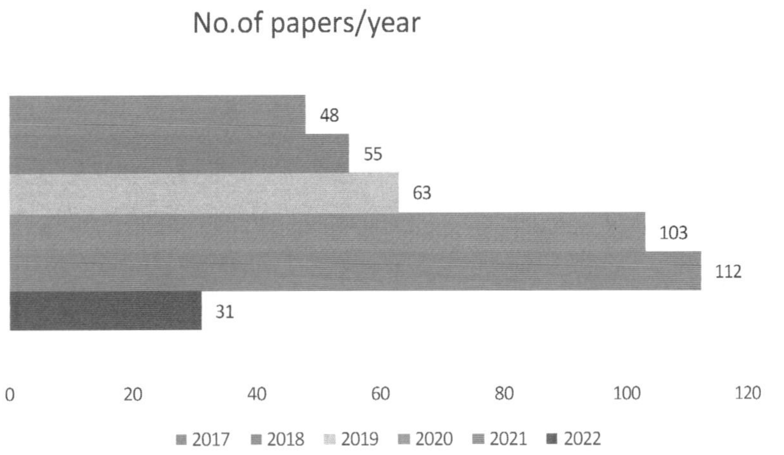

Figure 1. Bibliometric data.

The statistics were done using the data obtained from Web of Science, using a single criterion, key words such as: plastic, wastes and building materials. This topic has a great importance all over the word and is an area of new research studies.

2. Materials and Methods

It is well known that plastic is a recyclable material that can be reused in building materials. In this context, the present study aims to make masonry mortars reinforced with plastic fibers.

2.1. Materials

In the experimental program it has been proposed to develop three recipes of masonry mortars using as reinforcements plastic wastes, in a higher dosage as possible, in order to meet the current standards in terms of mechanical strength.

To observe the influence of plastic waste on mechanical strengths, of masonry mortars, the present study proposes the development of three types of masonry mortar recipes: non-reinforced masonry mortar class M10, masonry mortar reinforced with plastic fibers and masonry mortar reinforced with polypropylene fibers.

The materials used in the composition of the mortar are Portland cement, sand, water, plastic wastes fibers and polypropylene fibers.

The cement used in the recipes is Portland cement CEMI 52.5R type, a cement without any additions, with high initial mechanical strengths. According to the data sheet, the cement is composed from clincher (95-100%) and auxiliary components (0-5%). The initial setting time is 45 minutes, the stability is maximum 10mm and the compressive strength at 28 day is 52.2 N/mm^2.

The aggregates used in the experimental program are river sand with the grain size between 0-8 mm. The aggregates were washed and dried, before the preparation of recipes.

The polypropylene fibers utilized have a length of 12 mm and the recommended dosage in the composition of concrete is between 1-1.5 kg/m^3.

The plastic fibers used for reinforcement were obtained from PET bottles, cut into strips with the following dimensions: length 40mm, width 10mm and thickness 1mm. The dosage of plastic fibers used in the recipes is 1 kg/m^3.

The water used in the proposed recipes is potable water.

The recipes are presented in the Table 1 below:

Table 1. Proposed recipes in the experimental program.

Materials	U.M	RECIPE 1	RECIPE 2	RECIPE 3
Cement CEM I 52.5R	Kg	323	323	323
Aggregates	Kg	1600	1600	1600
Water	L	305	295.71	304.29
Polypropylene fibers	Kg	-	-	1
Plastic fibers	Kg	-	1	-

The study was conducted on 3 mortar recipes:

- Recipe 1 - standard masonry mortar M10 based on: cement, sand with grain size 0-8mm and water.
- Recipe 2 - M10 masonry mortar on cement base was used as reinforcement percentage for 1m^3 concrete, 1kg of plastic strips; sand with grain size 0-8mm and water.
- Recipe 3 - M10 masonry mortar based on cement reinforced with polypropylene fiber, 0-8mm sand and water.

2.2. Methods

The study of the behavior of the mortars in the experimental program was carried out on both fresh and hardened mortar at 28 days of age.

As regards the study of the characteristics of the fresh mortars, the apparent density and the consistency of the mortars determined with the spreading mass were followed.

As regards the study of the characteristics of the mortars in the hardened state, the following characteristics were studied: apparent density, flexural strength, and compressive strength.

2.2.1. Apparent density for fresh mortars. According to the norms (SR EN 1015-6:2001 and SR EN 1015-6/A1:2001), after the homogenization the mortar is placed in a cylindrical vessel with the volume of V_v=1 liter. The vessel is filled with mortar up to halfway up the rim. After removing the extension frame, the excess mortar is removed and with a metal blade.

Weigh the mortar pot and obtain the mass of the mortar, and then it is calculated the apparent density using the equation (1):

$$\rho_a = \frac{m}{V_v}\left[\frac{kg}{m^3}\right] \quad (1)$$

2.2.2. Consistency for fresh mortars. Consistency is a measure of the fluidity or moisture of fresh mortar and indicates a measure of the deformability of fresh mortar when subjected to a certain type of stress. Following the preparation of the proposed recipes, the determination of the consistency was carried out using the mass spreader according to SR EN 1013-13. After applying the 15 vibrations (at a frequency of approximately 1 vibration per second), with the mortar form kneading table, the diameter of the mortar is measured in two perpendicular directions using the screwdriver. The average value of the consistency will be calculated as the average value of two measurements.

2.2.3. Apparent density of hardened mortars. The apparent density of the hardened mortar was determined as the ratio between the dry mass of the mortar samples after drying in the oven and the volume of the samples tested (40x40x160 mm). Six prisms were made for each type of mortar studied.
2.2.4. Bending tensile strength of hardened mortars. To obtain the average values of the bending tensile strength, the samples with dimensions of 40x40x160 mm were prepared. After stripping the samples were kept and tested according to the current European standard SR EN 1015-11:2002, until at age of 28 days. Testing of the samples was carried out with the apparatus for determining the bending tensile strength.
2.2.5. Compressive strength of hardened mortars. Determination of the compressive strength was carried out at the age of 28 days of the tested mortars. The apparatus used for the determination was a hydraulic press. The determination was carried out on the half prisms resulting from the determination of the flexural strength. Testing was carried out according to the European standard SR EN 1015-11:2002.

3. Results

The result obtained on the studied mortars are presented below.

Apparent density for fresh mortars

The apparent density determination was carried out at the time of preparation of the proposed recipes, so all the results are shown in the Table 2 below.

Table 2. Bulk density of fresh mortars.

	RECIPE 1	RECIPE 2	RECIPE 3
Apparent density [kg/m^3]	2055	2095	1987

According to the table above, the mortar with the lowest density is M10 mortar reinforced with polypropylene fibers. All the mortars made comply with the standards in force. The mortar made with plastic strips is the mortar with the highest average value of apparent density.

3.1. Consistency for fresh mortars

The results obtained are shown in the Table 3 below.

Table 3. Consistency of fresh mortars.

	RECIPE 1	RECIPE 2	RECIPE 3
	Average value [mm]		
Consistency of fresh mortars	188.5	180	180

From the table above it can be seen that RECIPE1 mortar is the mortar with the highest consistency, which is the mortar with no reinforcement (figure 2). Regarding the results achieved for dispersed reinforced mortars, either with polypropylene fibers or plastic fibers, the average values obtained were the similar.

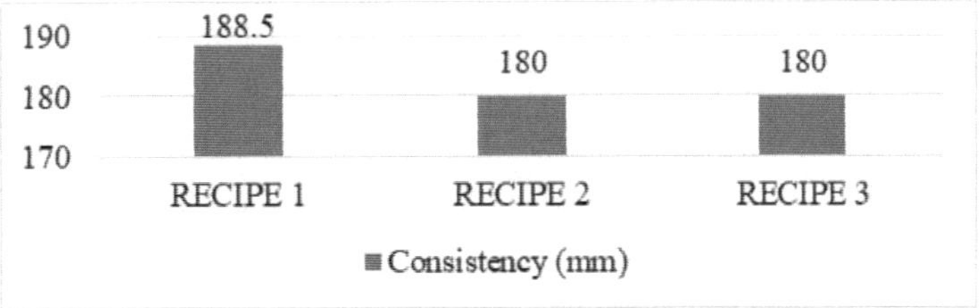

Figure 2. Consistency of fresh mortars.

For mortars with a density greater than 1200 kg/m^3, the spread should be 175±10mm. From the graph above, M10 mortars reinforced with plastic waste fall within the values given by the standard.

Non-reinforced M10 mortar has a slight increase average value compared to the values of the norms. M10 mortar - reinforced with polypropylene fibers also falls within the standard values.

3.2. Apparent density of hardened mortars

The average results obtained on the hardened mortars were determined at the age of 28 days and are shown in Table 4.

Table 4. Average value of apparent density of hardened mortars.

	RECIPE 1	RECIPE 2	RECIPE 3
	Average value [kg/m^3]		
Apparent density	2051	2135	2074

According to the table above, the lowest density hardened mortar is the unreinforced M10 mortar. The mortar made with plastic strips is the mortar with the highest average bulk density values. All the mortars made comply with the norms in force and must be between 1950 and 2200 kg/m^3.

3.3. Bending tensile strength of hardened mortars.

The results are shown in the table below. Testing was performed on 6 test specimens of 40x40x160mm at the age of 28 days, Table 5.

Table 5. Average value of bending tensile strength of hardened mortars.

	RECIPE 1	RECIPE 2	RECIPE 3
	Average value [N/mm^2]		
Bending tensile strength	3.953	4.257	3.718

The average value obtained for mortars made with plastic strips is 4.257 N/mm^2, which is the highest value obtained compared to the other 2 types of mortars studied. The unreinforced M10 mortar has a flexural strength of 3.953 N/mm^2 followed by the polypropylene fiber reinforced M10 mortar with a value of 3.718 N/mm^2.

3.4. Compressive strength of hardened mortars.

The results are shown in the table below. Testing was performed on 12 test specimens at the age of 28 days, Table 6.

Table 6. Average value of compressive strength of hardened mortars.

	RECIPE 1	RECIPE 2	RECIPE 3
	Average value [N/mm^2]		
Compressive strength	15.39	29.29	22.58

In terms of compressive strength, the mortar reinforced with plastic strips shows the highest values of compressive strength - 29.29 N/mm^2. The M10 mortar reinforced with polypropylene fibers has an average compressive strength of 22.58 N/mm^2, which is 1.29% lower than that of the mortar with plastic strips. The non-reinforced mortar has the lowest average compressive strength of 15.39 N/mm^2. In Figure 3, are presented the results obtained in terms of mechanical behavior.

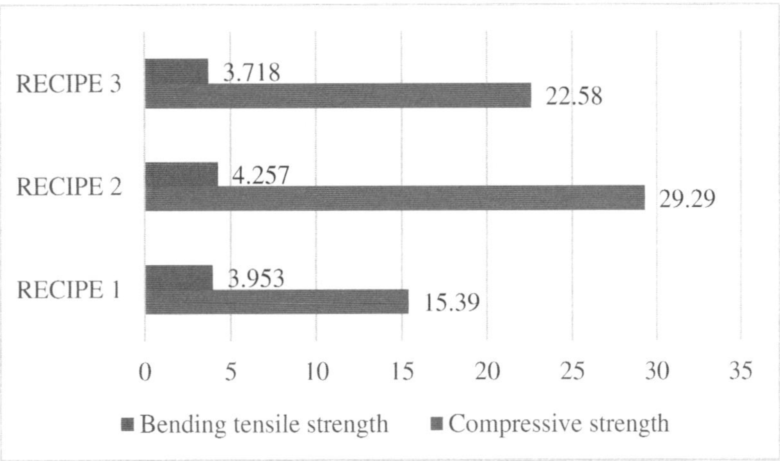

Figure 3. Comparative graphic on mechanical strength.

4. Conclusion

This paper addresses a topic of great global interest, as plastic waste is known to be an environmental hazard.

In view of sustainable development, this paper proposes the reuse of plastic waste in the production of masonry mortars, which are used as reinforcement for these mortars.

The results obtained showed a better performance of mortars reinforced with plastic strips compared to the other 2 types of mortars studied. Both in terms of flexural strength and compressive strength, these types reinforced with plastic strips show the highest values. It can be mentioned that this type of reinforcement helps to increase mechanical strengths compared to non-reinforced mortars.

The present study is a first step to follow the behavior of these types of materials, and in the future other types of recipes will be proposed to follow the behaviour of mortars with different percentages of reinforcement.

Acknowledgment

This paper was financially supported by the Project "Entrepreneurial competences and excellence research in doctoral and postdoctoral programs - ANTREDOC", project co-funded by the European Social Fund financing agreement no. 56437/24.07.2019

References

[1] https://ec.europa.eu/environment/topics/waste-and-recycling_ro, accesed on 10.04.2022.
[2] Directive (EU) 2019/904 of the European Parliament and of the Council, of 5 June 2019 on the reduction of the impact of certain plastic products on the environment.
[3] Precious Plastic just released a brick made from 100% recycled plastic to empower people to build constructions just like with LEGO (but a whole lot bigger), https://designwanted.com/architecture/precious-plastic-brick/, accessed on 06.05.2021.
[4] Haque Md S 2019 *Environmental Science and Pollution Research* **26**(36) 36163–36183.
[5] Nabajyoti S and De Brito J 2012 *Construction and Building Materials* **34** 385–401.
[6] Awoyera P O and Adesina A 2020 *Case Studies in Construction Materials* **12.**
[7] Foti D 2019 *Woodhead Publishing Series in Composites Science and Engineering* pp 387–410.
[8] Nergis, DDB, Vizureanu P, Sandu AV, Nergis, DPB and Bejinariu C 2022 *Materials* **15(1)** 202. DOI: 10.3390/ma15010202.
[9] Yahya Z, Abdullah MMAB, Ramli NM, Burduhos-Nergis DD and Abd Razak R 2018 *IOP Conf. Ser.: Mater. Sci. Eng.* **374** 012068. DOI: 10.1088/1757-899X/374/1/012068.
[10] Burduhos-Nergis D-P, Sandu A-V, Burduhos-Nergis D-D, Darabont D-C, Comaneci R-I and Bejinariu C 2019 *MATEC Web of Conferences* **290** 12004

[11] Vizureanu P and Agop M 2007 *Materials Transactions* **48(11)** 3021-3023. DOI: 10.2320/matertrans.MRP2007118.
[12] Ciubară A, Burlea SL, Axinte M, Cimpoeşu R, Chicet DL, Manole V, Burlea G and Cimpoeşu N 2018 *IOP Conference Series: Materials Science and Engineering* **444(3)** 032013.
[13] Javadi Toghchi M, Campagne C, Cayla A, Bruniaux P, Loghin C, Cristian I, Burgnies L and Chen Y 2019 *Synthetic Metals* **251** 75-84. doi: 10.1016/j.synthmet.2019.03.026.

AUTHOR INDEX